Test Item File
to accompany

BASIC STATISTICS

Stephen B. Jarrell

Prepared by
Marcel Maupin
Oklahoma State University

Wm. C. Brown Publishers
Dubuque, Iowa • Melbourne, Australia • Oxford, England

Contents

INTRODUCTION

This manual contains a printout of the test items and algorithm formulas from
ESATEST III that accompany *Basic Statistics*, 1e, by Stephen B. Jarrell. **ESATEST III** is
a computerized test preparation system. With **ESATEST III**, you can quickly and easily
create and print a test containing any combination of questions from the set of test
questions or algorithm formulas accompanying your textbook.

ESATEST III lets you choose test questions in five ways:
- by chapter and question number
- by viewing each question on the screen
- by criteria (according to difficulty level, learning objective, question type, or other criteria)
- by algorithm formulas
- by entering your own questions

With **ESATEST III** you can:

- Add your own questions and answers of any length or type (multiple choice, true/false, matching, essay, etc.);
- Produce complex mathematical formulas;
- Maintain separate tests and chapter questions for different instructors or class sections;
- Edit the questions/answers already provided;
- Create a test covering one or more chapters of the book;
- Create and edit your own graphics;
- Print scrambled versions of the same test;
- Include subscripts, superscripts, special characters, and symbols in the test questions.

ESATEST III provides easy-to-follow, context-sensitive Help screens throughout the
program. Commands, available through the keyboard or a mouse, are listed in the pull-
down menus and on-screen prompts so that even the new user can create, modify, and
print tests with ease.

The **ESAPAINT** graphics program, included with **ESATEST III** packages that require
graphics, lets you use a mouse to alter illustrations provided in your chapter disks, add
text labels, or create graphs and other figures of your own.

ESAGRADE is designed to assist educators in recording and processing results from
student examinations and assignments, and then provide a detailed statistical analysis of
each student or class.

ESAGRADE can:
- Set up new classes.
- Record grades from tests, quizzes, and assignments.
- Automatically grade tests prepared by ESATEST III using SCANTRON.
- Analyze grades and produce class and student statistics.
- Provide graphic display of statistics.

ESAGRADE combines a powerful database with powerful analytical capabilities. A full set of statistics may be generated including:

- Class Average
- Class Standard Deviation
- Class Curve

- Student Average For Tests
- Student Average For Assignments
- Student Weighted Average

ESAGRADE is easier to use and much more efficient than a traditional gradebook. Test results, cumulative average, class standings, etc., are available at the touch of a keyboard or the click of a mouse.

Complete documentation can be found in your **ESATEST III** package. For additional information, please contact your Wm. C. Brown Publishers representative at 1-800-258-2385.

Algorithm Formulas

This is a list of operators used in algorithm formulas, and the syntax of the formulas.

Some Helpful Abbreviations

UNARY OPERATORS

Function	Description	Domain
sin	sine	
cos	cosine	
tan	tangent	
asin	arcsine	$-1 \leq x \leq = 1$
acos	arccosine	$-1 \leq x \leq 1$
atan	arctangent	
sinh	hyperbolic sine	
cosh	hyperholic cosine	
tanh	hyperholic tangent	
e	exponential function	
ln	natural logarithm	$x > 0$
log	logarithm to the base 10	$x > 0$
cei	ceiling x, (the smallest integer not less than x) e.g., $x = 1.02$, ceiling $x = 2$ $x = -1.02$, ceiling $x = -1$	
flr	floor x, (the largest integer not greater than x) e.g., $x = 1.02$, floor $x = 1$ $x = -1.02$, floor $x = -2$	

BINARY OPERATORS

Function	Description
+	Addition
-	Subtraction
*	Multiplication
/	Division denominator is not 0
^	Power $x^{\wedge}y$, except: $x = 0$, $y \leq 0$, $x < 0$, y not an integer

CHAPTER 1

TRUE/FALSE

1.Descriptive statistics uses inferences.

   Answer: F Type: T   Sect: 1 Obj: 2   RANDOM: Y

2.Statistics is divided into two main areas called descriptive and hypothetical statistics.

   Answer: F Type: T   Sect: 1 Obj: 2   RANDOM: Y

3.A population is the totality of all subjects possessing certain common characteristics that are being studied.

   Answer: T Type: T   Sect: 1 Obj: 1   RANDOM: Y

4.A sportswriter lists the winning times of all runners in the Boston Marathon. This would be an example of descriptive statistics.

   Answer: T Type: T   Sect: 1 Obj: 2   RANDOM: Y

5.A medical researcher tests a new drug and predicts that it will be beneficial to 60% of those who use it. This is an example of descriptive statistics.

   Answer: F Type: T   Sect: 1 Obj: 2   RANDOM: Y

6.The variable "sex" would be an example of a qualitative variable.

   Answer: T Type: T   Sect: 2 Obj: 3   RANDOM: Y

7.The number of hot dogs sold by a street vendor would be an example of a qualitative variable.

   Answer: F Type: T   Sect: 2 Obj: 3   RANDOM: Y

8.The amount of time it takes for a light bulb to burn out would be an example of a quantitative variable.

   Answer: T Type: T   Sect: 2 Obj: 3   RANDOM: Y

9.If a recorded height is 61 inches, this is an example of a quantitative variable.

   Answer: T Type: T   Sect: 2 Obj: 3   RANDOM: Y

10. The subject matter that a college professor teaches would be an example of ordinal level data.

    Answer: F Type: T   Sect: 2 Obj: 4   RANDOM: Y

11. An example of interval level data would be IQ scores.

    Answer: T Type: T   Sect: 2 Obj: 4   RANDOM: Y

MULTIPLE CHOICE

12. The collection, organization, summarization, and presentation of data is called
    a. inferential statistics
    b. a population
    c. probability
    d. descriptive statistics

    Answer: d Type: M   Sect: 1 Obj: 2   RANDOM: Y

13. When the values of a variable are collected, what are they called?
    a. random variables
    b. data
    c. a sample
    d. a population

    Answer: b Type: M   Sect: 1 Obj: 1   RANDOM: Y

14. Which of the following is an area of inferential statistics?
    a. collecting data
    b. organizing data
    c. generalizing from samples to populations
    d. generalizing from populations to samples

    Answer: c Type: M   Sect: 1 Obj: 2   RANDOM: Y

15. The number of pounds of garbage generated by people in Great Britain would be an example of what type of variable?
    a. quantitative
    b. random
    c. qualitative
    d. homogeneous

    Answer: a Type: M   Sect: 2 Obj: 3   RANDOM: Y

16. A person's blood type would be an example of what type of data?
    a. descriptive
    b. quantitative
    c. qualitative
    d. homogeneous

    Answer: c Type: M   Sect: 2 Obj: 3   RANDOM: Y

17. A variable that can be placed into distinct categories according to some characteristic
    but cannot be ranked is called a
    a. continuous variable
    b. qualitative variable
    c. quantitative variable
    d. homogeneous variable

    Answer: b Type: M   Sect: 2 Obj: 3   RANDOM: Y

18. What is a level of measurement which consists of classifying data into mutually
    exclusive categories in which no order or ranking can be imposed?
    a. nominal level
    b. ordinal level
    c. interval level
    d. ratio level

    Answer: a Type: M   Sect: 2 Obj: 4   RANDOM: Y

19. The salaries of auto workers would be an example of what type of data?
    a. nominal level
    b. ordinal level
    c. interval level
    d. ratio level

    Answer: d Type: M   Sect: 2 Obj: 4   RANDOM: Y

20. A person's political affiliation would be an example of what type of data?
    a. nominal level
    b. ordinal level
    c. interval level
    d. ratio level

    Answer: a Type: M   Sect: 2 Obj: 4   RANDOM: Y

21. A sample in which every possible set of items has an equal chance of selection is called
    what type of sample?
    a. probability   b. random
    c. representative   d. stratified

    Answer: b Type: M   Sect: 1 Obj: 1   RANDOM: Y

22. When evaluating arithmetic expressions requiring the use of levels of operation, what operations are performed first?
    a. multiplication and division
    b. addition and subtraction
    c. all exponential and logarithmic operations
    d. all statistical operations

    Answer: c Type: M   Sect: 3 Obj: 5   RANDOM: Y

SHORT ANSWER

Directions for the problems below: Identify the following statements as inferential descriptive statistics.

23. 60% of the students taking college algebra will pass.

    Answer: descriptive Type: S   Sect: 1 Obj: 2   RANDOM: Y

24. Based on past figures, it is predicted that only 52% of registered voters will vote in the next election.

    Answer: inferential Type: S   Sect: 1 Obj: 2   RANDOM: Y

25. John got a better grade on his second chemistry test because he studied harder for it.

    Answer: inferential Type: S   Sect: 1 Obj: 2   RANDOM: Y

26. There is a relationship between smoking and lung cancer.

    Answer: inferential Type: S   Sect: 1 Obj: 2   RANDOM: Y

27. During the Memorial Day weekend, 20 people were killed in car accidents in Washington, D.C.

    Answer: descriptive Type: S   Sect: 1 Obj: 2   RANDOM: Y

28. Of the 30 nurses in pediatrics, 10 work at night and 20 work during the day.

    Answer: descriptive Type: S   Sect: 1 Obj: 2   RANDOM: Y

29. Gilbert Johnston got 350 votes for county clerk and Anna Henson got 460 votes for county clerk.

    Answer: descriptive Type: S   Sect: 1 Obj: 2   RANDOM: Y

30. An economist estimates the average income of people in Cheyenne, Wyoming, to be $19,750.

    Answer: inferential Type: S   Sect: 1 Obj: 2   RANDOM: Y

31. Vicky bowled 5 games and found that her average was 136.

    Answer: descriptive Type: S   Sect: 1 Obj: 2   RANDOM: Y

32. Jerry had a 76% average in Calculus I. Based on this data, he estimates that he will have a 78% average in Calculus II.

    Answer: inferential Type: S   Sect: 1 Obj: 2   RANDOM: Y

33. Joyce had grades of 68, 72, 81, 75, and 98 on her five exams in history. Based on this data, the highest grade Joyce received was a 98.

    Answer: descriptive Type: S   Sect: 1 Obj: 2   RANDOM: Y

34. Joyce had grades of 68, 72, 81, 75 and 98 on her five exams in history. Based on this data, the reason Joyce did so well on the last test was that it dealt with modern-day history and Joyce was very knowledgeable about this subject.

    Answer: inferential Type: S   Sect: 1 Obj: 2   RANDOM: Y

35. Based on last year's electric bills, Mr. Smith estimates that it will cost about $85 a month for his electricity this year.

    Answer: inferential Type: S   Sect: 1 Obj: 2   RANDOM: Y

Directions for the problems below: Identify the following as quantitative or qualitative variables.

36. the number of pages typed by a secretary

    Answer: quantitative Type: S   Sect: 2 Obj: 3   RANDOM: Y

37. the heights of pro basketball players

    Answer: quantitative Type: S   Sect: 2 Obj: 3   RANDOM: Y

38. the religious affiliation of sophomores in high school

    Answer: qualitative Type: S   Sect: 2 Obj: 3   RANDOM: Y

39. the length of time it takes to do an accounting problem

    Answer: quantitative Type: S Sect: 2 Obj: 3   RANDOM: Y

40. the number of children in a family

    Answer: quantitative Type: S Sect: 2 Obj: 3   RANDOM: Y

41. the weights of elephants in zoos across the United States

    Answer: quantitative Type: S Sect: 2 Obj: 3   RANDOM: Y

42. the favorite flavor of ice cream of first graders

    Answer: qualitative Type: S Sect: 2 Obj: 0   RANDOM: Y

43. the number of golf balls lost each day at the country club

    Answer: quantitative Type: S Sect: 2 Obj: 3   RANDOM: Y

44. the favorite color of car of the faculty at Smalltown College

    Answer: qualitative Type: S Sect: 2 Obj: 3   RANDOM: Y

45. the number of miles driven by American Van Lines each day

    Answer: quantitative Type: S Sect: 2 Obj: 3   RANDOM: Y

46. the effectiveness of a speaker at a public speaking contest

    Answer: qualitative Type: S Sect: 2 Obj: 3   RANDOM: Y

47. the number of sand traps at a golf course

    Answer: quantitative Type: S Sect: 2 Obj: 3   RANDOM: Y

48. the amount of time it takes to run a 20-mile race

    Answer: quantitative Type: S Sect: 2 Obj: 3   RANDOM: Y

49. the political party preference of college seniors

    Answer: qualitative Type: S Sect: 2 Obj: 3   RANDOM: Y

50. the number of apartments for rent at Glendale Estates

Answer: quantitative Type: S   Sect: 2 Obj: 3   RANDOM: Y

Directions for the problems below: Evaluate the following expressions.

51. $60 \cdot 2 + 5(7)^2$

Answer: 365 Type: S   Sect: 3 Obj: 5   RANDOM: Y

52. $100 - (5 \div 10)^2 + 20 \div 4 \cdot 5$

Answer: 124.75 Type: S   Sect: 3 Obj: 5   RANDOM: Y

53. $(6 - 3)^2 - 8^2 \div 4$

Answer: -7 Type: S   Sect: 3 Obj: 5   RANDOM: Y

54. $75 - 6\sqrt{100} \div 15 + 7 \cdot 9$

Answer: 134 Type: S   Sect: 3 Obj: 5   RANDOM: Y

55. $14 + 2(3)^3 + \left[\sqrt{400} - 7\right]^2$

Answer: 237 Type: S   Sect: 3 Obj: 5   RANDOM: Y

56. $200 - 36 \div 6 \cdot 6 - 6 + \sqrt{49}$

Answer: 165 Type: S   Sect: 3 Obj: 5   RANDOM: Y

Classify each of the following as nominal level data, ordinal level data, interval level data, or ratio level data.

57. the marital status of people in Las Vegas

Answer: nominal level Type: S   Sect: 2 Obj: 4   RANDOM: Y

58. ratings of restaurants in New York City (poor, fair, good, excellent)

Answer: ordinal level Type: S   Sect: 2 Obj: 4   RANDOM: Y

59. the time required by students in COBOL I to finish a computer program

    Answer: ratio level Type: S   Sect: 2 Obj: 4   RANDOM: Y

60. SAT scores of students at Berkeley High School

    Answer: interval level Type: S   Sect: 2 Obj: 4   RANDOM: Y

61. world rankings of heavyweight boxers

    Answer: ordinal level Type: S   Sect: 2 Obj: 4   RANDOM: Y

62. the hair color of students in English 101

    Answer: nominal level Type: S   Sect: 2 Obj: 4   RANDOM: Y

63. the ages of employees working for Lockheed

    Answer: ratio level Type: S   Sect: 2 Obj: 4   RANDOM: Y

64. temperatures of various fish tanks at a large pet store

    Answer: interval level Type: S   Sect: 2 Obj: 4   RANDOM: Y

65. eggs are rated as small, medium, large, or jumbo

    Answer: ordinal level Type: S   Sect: 2 Obj: 4   RANDOM: Y

66. the zip code of alumni of State University

    Answer: nominal level Type: S   Sect: 2 Obj: 4   RANDOM: Y

67. the classification of books into groups by a university library

    Answer: nominal level Type: S   Sect: 2 Obj: 4   RANDOM: Y

68. the assignment of ratings at a music contest--I for excellent, II for good, and III for fair

    Answer: ordinal level Type: S   Sect: 2 Obj: 4   RANDOM: Y

69. the measurement of the weights of sumo wrestlers

    Answer: ratio level Type: S   Sect: 2 Obj: 4   RANDOM: Y

70. the outside temperature measured in Celsius

    Answer: interval level Type: S   Sect: 2 Obj: 4   RANDOM: Y

71. the measurement in cubic feet of the volume of crude oil in an oil tanker

    Answer: ratio level Type: S   Sect: 2 Obj: 4   RANDOM: Y

72. What is a statistic?

    Answer: A number that summarizes information in a data set. Type: S   Sect: 1

    Obj: 1   RANDOM: Y

73. What is the type of sampling called that results in the items sampled no longer being useful?

    Answer: destructive sampling Type: S   Sect: 1 Obj: 1   RANDOM: Y

74. What is a value that does not vary with each observation called?

    Answer: constant Type: S   Sect: 3 Obj: 1   RANDOM: Y

75. What is a subgroup or subset of the population called?

    Answer: a sample Type: S   Sect: 2 Obj: 1   RANDOM: Y

76. A large university wants to estimate salaries. They select 50 faculty members and use their salaries as a basis for the whole university. What term is used to designate these 50 faculty members?

    Answer: a sample   Type: S   Sect: 1 Obj: 1   RANDOM: Y

77. The number of hours of television watched by fifth graders at Westport Elementary School would be an example of what type of data?

    Answer: quantitative Type: S   Sect: 3 Obj: 3   RANDOM: Y

78. What are the two branches of statistics?

    Answer: descriptive and inferential Type: S   Sect: 1 Obj: 2   RANDOM: Y

79. The number of books in the Wilmont Library would be an example of what type of data?

    Answer: quantitative data Type: S   Sect: 2 Obj: 3   RANDOM: Y

80. The geographic location of a factory would be an example of what type of data?

   Answer: qualitative data Type: S   Sect: 2 Obj: 3   RANDOM: Y

Directions for the problems below: Evaluate the following expressions.

81. $16 - 2 \cdot 3 + 5 \cdot 4$

   Answer: 30 Type: S   Sect: 3 Obj: 5   RANDOM: Y

82. $\left[100 - 5\sqrt{64}\right]^2 - 9^2$

   Answer: 3519 Type: S   Sect: 3 Obj: 5   RANDOM: Y

83. $8^2 - 7 + 3 \cdot 9 - \sqrt{121}$

   Answer: 73 Type: S   Sect: 3 Obj: 5   RANDOM: Y

84. $64 \div 8 \cdot 2^2 + 5\sqrt{16} - 7 + 2 \cdot 6$

   Answer: 57 Type: S   Sect: 3 Obj: 5   RANDOM: Y

85. $18^2 - 3(7 - 5)^3 + \sqrt{4^2 + 3^2}$

   Answer: 305 Type: S   Sect: 3 Obj: 5   RANDOM: Y

86. Calculate: $\displaystyle\sum_{i=1}^{6} 3i$

   Answer: 63 Type: S   Sect: 3 Obj: 5   RANDOM: Y

87. Calculate: $\displaystyle\sum_{i=1}^{7} (i^2 + 3)$

   Answer: 161 Type: S   Sect: 3 Obj: 5   RANDOM: Y

88.
   Calculate: $\displaystyle\sum_{i=1}^{10} 4$

   Answer: 40 Type: S   Sect: 3 Obj: 5   RANDOM: Y

Use the following information to solve the problems below.

$X_1 = 1$	$Y_1 = 2$	$Z_1 = -2$
$X_2 = 3$	$Y_2 = 0$	$Z_2 = 5$
$X_3 = 4$	$Y_3 = -1$	$Z_3 = 1$
$X_4 = 0$	$Y_4 = 6$	$Z_4 = 3$
$X_5 = -2$	$Y_5 = 2$	$Z_5 = 0$

89.
   Calculate: $\displaystyle\sum_{i=1}^{5} XiYi$

   Answer: -6 Type: S   Sect: 3 Obj: 5   RANDOM: Y

90.
   Calculate: $\displaystyle\left[\sum_{i=1}^{5} Zi\right]^2$

   Answer: 49 Type: S   Sect: 3 Obj: 5   RANDOM: Y

91.
   Calculate: $\displaystyle\sum_{i=1}^{5} (Zi - yi)^2$

   Answer: 58 Type: S   Sect: 3 Obj: 5   RANDOM: Y

92.
   Calculate: $\displaystyle\sum_{i=1}^{5} (Xi)^2$

   Answer: 30 Type: S   Sect: 3 Obj: 5   RANDOM: Y

93.
   Calculate: $\displaystyle\left[\sum_{i=1}^{5} yi\right]^2 - \sum_{i=1}^{5} (yi)^2$

   Answer: 36 Type: S   Sect: 3 Obj: 5   RANDOM: Y

94.
Calculate: $\displaystyle\sum_{i=1}^{5} [(Zi)^2 - Xi]$

Answer: 33 Type: S   Sect: 3 Obj: 5   RANDOM: Y

95.
Calculate: $\displaystyle\sum_{i=1}^{5} (XiYiZi)$

Answer: -8 Type: S   Sect: 3 Obj: 5   RANDOM: Y

CHAPTER 2

TRUE/FALSE

1.When data are collected in original form, they are called raw data.

Answer: T Type: T Sect: 1 Obj: 1 RANDOM: Y

2.Sandy is constructing a frequency distribution of blood types. She has 20 pieces of data, and 8 of the 20 are blood type O. The percentage of those with type O blood would be 30%.

Answer: F Type: T Sect: 2 Obj: 1 RANDOM: Y

3.For the data set 5, 9, 12, 11, 2, 4, 9, and 8, the range would be 10.

Answer: T Type: T Sect: 1 Obj: 1 RANDOM: Y

4.The class limits should have one additional decimal place value more than the data.

Answer: F Type: T Sect: 1 Obj: 1 RANDOM: Y

5.If the class limits are 7.6–7.9, the boundaries for that class would be 7.5–8.

Answer: F Type: T Sect: 1 Obj: 1 RANDOM: Y

6.In the class 13.2–15.6, the midpoint would be 14.4.

Answer: T Type: T Sect: 1 Obj: 1 RANDOM: Y

7.The first two classes of a distribution are 5–9 and 10–14. The class size would be 4.

Answer: F Type: T Sect: 1 Obj: 1 RANDOM: Y

8.In the class 25–30, the upper class boundary would be 30.5.

Answer: T Type: T Sect: 1 Obj: 1 RANDOM: Y

9.In the class 24–34, the lower class limit would be 24.

Answer: T Type: T Sect: 1 Obj: 1 RANDOM: Y

10.A histogram is a graph that displays data using vertical bars of various heights to represent the frequencies.

Answer: T Type: T Sect: 5 Obj: 2 RANDOM: Y

11. The frequency polygon is a graph that displays data using lines to connect the lower class limits.

    Answer: F Type: T   Sect: 4 Obj: 2   RANDOM: Y

12. An ogive uses vertical or horizontal bars to represent the frequencies of a distribution.

    Answer: F Type: T   Sect: 6 Obj: 2   RANDOM: Y

MULTIPLE CHOICE

13. What is not an advantage of a stem-and-leaf display?
    a. It organizes the data.
    b. It retains information about the actual values.
    c. It communicates the pattern or distribution of the data.
    d. It depicts a line graph of the data.

    Answer: d Type: M   Sect: 7 Obj: 2   RANDOM: Y

14. The graph of a cumulative frequency distribution is called
    a. an ogive   b. a histogram
    c. a frequency polygon   d. a stem-and-leaf display

    Answer: a Type: M   Sect: 6 Obj: 2   RANDOM: Y

15. A tabular display of values that shows the number, proportion, or percentage of values in the data set that are larger or smaller than certain values is called a
    a. relative frequency distribution
    b. cumulative frequency distribution
    c. histogram
    d. frequency distribution

    Answer: b Type: M   Sect: 3 Obj: 2   RANDOM: Y

16. Another name for an extreme value of a data set is
    a. a sample   b. a population
    c. an outlier   d. a mode

    Answer: c Type: M   Sect: 1 Obj: 1   RANDOM: Y

17. For the data 5, 9, 2, 11, 7, 19, and 8, what is the range?
   a. 4
   b. 3
   c. 14
   d. 17

   Answer: d Type: M   Sect: 1 Obj: 1   RANDOM: Y

18. What would be the class boundaries for the class 3.52–3.58?
   a. 3.515–3.585
   b. 3.500–3.600
   c. 3.525–3.585
   d. 3.510–3.590

   Answer: a Type: M   Sect: 1 Obj: 1   RANDOM: Y

19. What is the midpoint of the class 20.01–21.06?
   a. 20.01   b. 21.06   c. 1.05   d. 20.535

   Answer: d Type: M   Sect: 1 Obj: 1   RANDOM: Y

20. What is the midpoint of the class 13.2–16.4?
   a. 13.2
   b. 14.8
   c. 16.4
   d. 14.3

   Answer: b Type: M   Sect: 1 Obj: 1   RANDOM: Y

21. In a frequency distribution, two classes are 40–46 and 47–53. What is the class size?
   a. 6
   b. 8
   c. 7
   d. 5

   Answer: c Type: M   Sect: 1 Obj: 1   RANDOM: Y

22. In the class 20–28, what is the lower class limit?
   a. 20
   b. 19
   c. 19.5
   d. 20.5

   Answer: a Type: M   Sect: 1 Obj: 1   RANDOM: Y

23. Using the class 37–45, what is the upper class boundary?
    a. 45
    b. 45.5
    c. 44.5
    d. 46

    Answer: b Type: M   Sect: 1 Obj: 1   RANDOM: Y

24. For the class 17.1–19.2, what is the class size?

    a. 2.1   b. 2.2   c. 17.1   d. 18.15

    Answer: b Type: M   Sect: 1 Obj: 1   RANDOM: Y

25. Using the class 17.25–18.10, what is the upper class limit?
    a. 18.20
    b. 18.10
    c. 18.105
    d. 18.11

    Answer: b Type: M   Sect: 2 Obj: 1   RANDOM: Y

26. For the class 19–35, what is the class midpoint?

    a. 19   b. 35   c. 27   d. 16

    Answer: c Type: M   Sect: 1 Obj: 1   RANDOM: Y

27. The following frequency distribution was obtained.

Degrees	Days
18–22	5
23–27	7
28–32	10
33–37	12
28–42	13

    How many pieces of data were less than 33?
    a. 12
    b. 13
    c. 22
    d. 34

    Answer: c Type: M   Sect: 3 Obj: 2   RANDOM: Y

28. Consider the following frequency distribution of the number of votes for mayor in different precincts:

Votes	Number of Precints
0–99	3
100–199	10
200–299	25
300–399	15
400–499	22

What proportion of the votes are between 200 and 299?
a. 25　　　　　b. 1/3　　　　　c. 1/5　　　　　d. 1/4

Answer: b Type: M   Sect: 3 Obj: 1   RANDOM: Y

29. For the class 110–117, what is the lower class boundary?

a. 110　　　　　b. 117　　　　　c. 117.5　　　　　d. 109.5

Answer: d Type: M   Sect: 1 Obj: 1   RANDOM: Y

30. For the class 250–260, what is the upper class limit?

a. 250　　　　　b. 260　　　　　c. 260.5　　　　　d. 10

Answer: b Type: M   Sect: 1 Obj: 1   RANDOM: Y

SHORT ANSWER

Directions for the problems below: Find the class boundaries.

31. 8–18

Answer: 7.5–18.5 Type: S   Sect: 1 Obj: 1   RANDOM: Y

32. 120–126

Answer: 119.5–126.5 Type: S   Sect: 1 Obj: 1   RANDOM: Y

33. 13.2–13.8

Answer: 13.15–13.85 Type: S   Sect: 1 Obj: 1   RANDOM: Y

34.4.15–4.19

   Answer: 4.145–4.195 Type: S   Sect: 2 Obj: 1   RANDOM: Y

35.10.2–10.3

   Answer: 10.15–10.35 Type: S   Sect: 2 Obj: 1   RANDOM: Y

Directions for the problems below: Find the class midpoint.

36.7–11

   Answer: 9 Type: S   Sect: 1 Obj: 1   RANDOM: Y

37.420–428

   Answer: 424 Type: S   Sect: 1 Obj: 1   RANDOM: Y

38.11.72–11.76

   Answer: 11.74 Type: S   Sect: 1 Obj: 1   RANDOM: Y

39.17.2–18.8

   Answer: 18 Type: S   Sect: 1 Obj: 1   RANDOM: Y

40.9.1–9.2

   Answer: 9.15 Type: S   Sect: 2 Obj: 1   RANDOM: Y

Directions for the problems below: Determine the class size.

41.18–22

   Answer: 5 Type: S   Sect: 2 Obj: 1   RANDOM: Y

42.100–109

   Answer: 10 Type: S   Sect: 2 Obj: 1   RANDOM: Y

43.12.25–12.31

   Answer: 0.07 Type: S   Sect: 2 Obj: 1   RANDOM: Y

44. 8.2–8.4

    Answer: 0.3 Type: S   Sect: 2   Obj: 1   RANDOM: Y

45. 9.8–11.6

    Answer: 1.9 Type: S   Sect: 2   Obj: 1   RANDOM: Y

46. 40–65

    Answer: 26 Type: S   Sect: 1   Obj: 1   RANDOM: Y

47. 19.2–21.3

    Answer: 2.2 Type: S   Sect: 1   Obj: 1   RANDOM: Y

48. In a survey of 24 different grocery stores, the number of different types of canned beans was collected. Construct a frequency distribution using 6 classes and beginning the first class with 10.

    15 20 30 16 10 12 18 31 15 11 19 35
    14 17 24 25 27 12 15 18 36 13 19 26

Answer:

Limits	f
10–14	6
15–19	9
20–24	2
25–29	3
30–34	2
35–39	2

Type: S   Sect: 1   Obj: 1   RANDOM: Y

49. The following is a list of the ages of 36 secretaries at a temporary employment agency. Construct a frequency distribution using 9 classes and starting with 18.

```
19   20   18   25   28   30   32   23   19   24   35
40   23   25   42   27   21   25   29   34   38   42
28   32   37   42   36   29   27   21   18   25   27
32   34   29
```

Answer:

Limits	f
18–20	5
21–23	4
24–26	5
27–29	8
30–32	4
33–35	3
36–38	3
39–41	1
42–44	3

Type: S   Sect: 1   Obj: 1   RANDOM: Y

50. A psychology teacher records the following grades on his 200 point final. Construct a frequency distribution using 10 classes starting with 88.

```
95   115   132   150   170   180   195   175   160   100    92
98   112   137   142   172   179   195   182   179   173   162
135   122   191   182   179   161   158   147   148   172   162
165   192   137
```

Answer:

Limits	f
88–98	3
99–109	1
110–120	2
121–131	1
132–142	5
143–153	3
154–164	5
165–175	6
176–186	6
187–197	4

Type: S   Sect: 1   Obj: 1   RANDOM: Y

51. The following data represent the scores on a history examination. Construct a frequency distribution using 6 classes beginning with 45.

53	65	90	78	95	82	72	88
58	62	81	75	83	70	63	56
89	92	73	72	66	89	98	83
49	72	77	85	87	90	58	65

Answer:

Scores	f
45–53	2
54–62	4
63–71	5
72–80	7
81–89	9
90–98	5

Type: S   Sect: 1 Obj: 1   RANDOM: Y

52. In a gymnastic competition, the following scores were recorded. Construct a frequency distribution using 9 classes beginning with 8.75.

9.65	9.5	10	9.54	9.84
9.4	9.95	9.68	8.95	9.91
8.75	9.35	9.40	9.65	9.75
9.85	9.9	10	9.55	9.64
9.85	9.35	9.15	9.4	9.25

Answer:

Limits	f
8.75–8.89	1
8.90–9.04	1
9.05–9.19	1
9.20–9.34	1
9.35–9.49	5
9.50–9.64	4
9.65–9.79	4
9.80–9.94	5
9.95–10.09	3

Type: S   Sect: 1 Obj: 1   RANDOM: Y

53.The weights of first graders at Anderson Elementary were recorded as follows. Construct
a frequency distribution using 6 classes and starting with 45.

45   50   60   65   48   52   62   57   62   70   51   59
65   49   46   50   51   53   58   62   67   71   59   64

Answer:

Limits	f
45–49	4
50–54	6
55–59	4
60–64	5
65–69	3
70–74	2

Type: S   Sect: 1   Obj: 1   RANDOM: Y

54.A frequency distribution is listed below:

Limits	f
46–52	8
53–59	10
60–66	12
67–73	7
74–80	2

What is the class size?

Answer: 7 Type: S   Sect: 1   Obj: 1   RANDOM: Y

55.A frequency distribution is listed below:

Limits	f
46–52	8
53–59	10
60–66	12
67–73	7
74–80	2

What is the midpoint of the class 60–66?

Answer: 63 Type: S   Sect: 1   Obj: 1   RANDOM: Y

56. A frequency distribution is listed below:

Limits	f
46–52	8
53–59	10
60–66	12
67–73	7
74–80	2

What is the lower class boundary of the class 74–80?

Answer: 73.5 Type: S   Sect: 1 Obj: 1   RANDOM: Y

57. A frequency distribution is listed below:

Limits	f
46–52	8
53–59	10
60–66	12
67–73	7
74–80	2

How many items are in this sample?

Answer: 39 Type: S   Sect: 1 Obj: 1   RANDOM: Y

58. If the data were in thousands, such as 15,000, 6,000, 7,000, etc., what would be the boundaries of a hypothetical class such as
6000–9000?

Answer: 5,500–9,500 Type: S   Sect: 1 Obj: 1   RANDOM: Y

59. If a frequency distribution had class boundaries 150.5–159.5, what would the class size be?

Answer: 9 Type: S   Sect: 1 Obj: 1   RANDOM: Y

60. A frequency distribution is listed below.

Scores	f
50–54	3
55–59	5
60–64	8
65–69	12
70–74	9
75–79	2

What is the upper class boundary of the class 55–59?

Answer: 59.5 Type: S   Sect: 1 Obj: 1   RANDOM: Y

61. Construct a histogram: The following data revealed the number of hours of television watched by 130 five-year-olds on a Saturday.

Class Limits	Frequency
0–2	15
3–5	40
6–8	45
9–11	20
12–14	10

Answer:

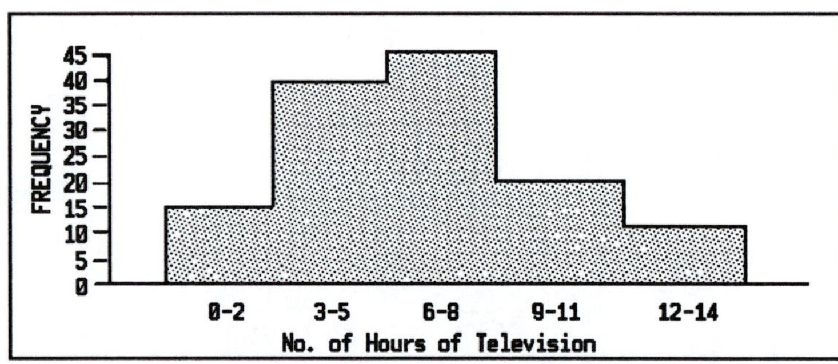

Type: S   Sect: 5 Obj: 2   RANDOM: Y

62. Construct a histogram: The tourism department in Aspen, Colorado, wanted to know the length of stay at their resorts. They collected the following data:

Length of Stay (days)	Frequency
3	8
4	10
5	15
6	20
7	28
8	18

Answer:

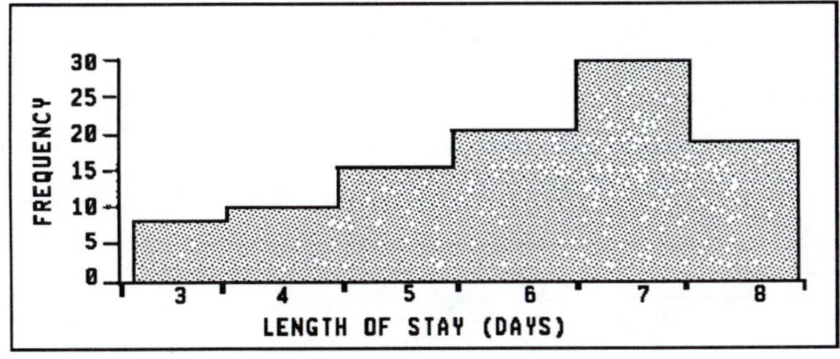

Type: S   Sect: 5   Obj: 2   RANDOM: Y

63. Construct a histogram: The following data indicate the monthly salaries of a sample of 70 nurses.

Salary	Frequency
1,900	5
2,200	10
2,500	20
2,800	25
3,100	8
3,400	2

Answer:

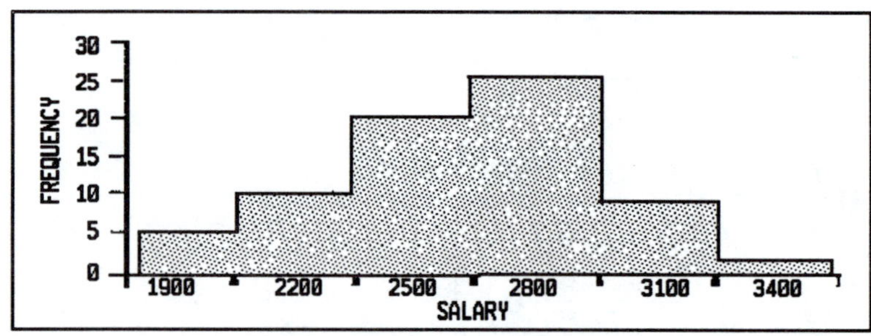

Type: S   Sect: 5   Obj: 2   RANDOM: Y

64. Construct a histogram: The following is the distribution of the number of mistakes 70 students made when translating an article from Spanish to English.

Class Limits	Frequency
0–4	18
5–9	29
10–14	13
15–19	8
20–24	2

Answer:

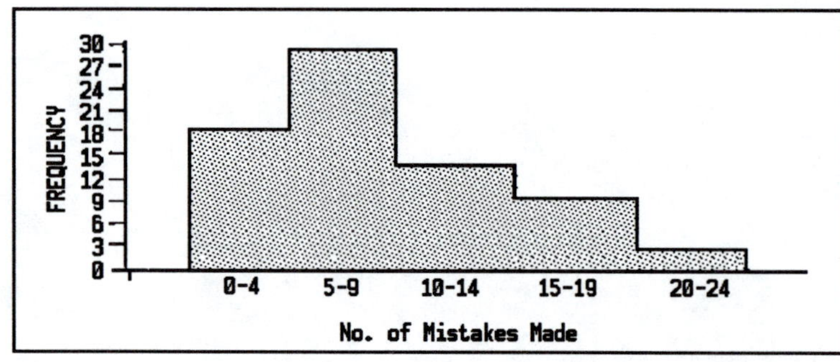

Type: S   Sect: 5   Obj: 2   RANDOM: Y

65. Construct a frequency polygon: The number of cups of coffee drunk by 150 employees at Sears were as follows:

No. of Cups	Frequency
0	15
1	20
2	30
3	45
4	25
5	15

Answer:

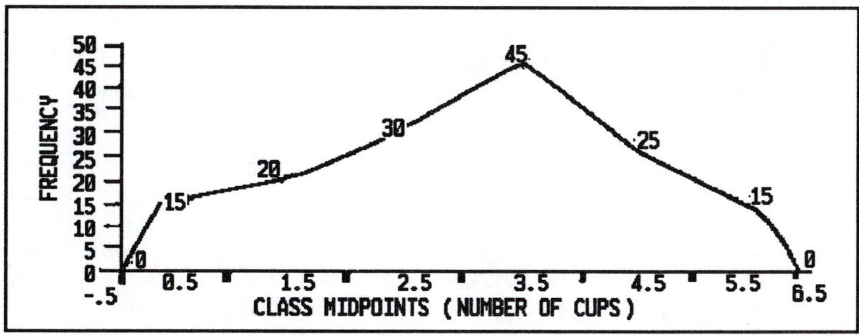

Type: S   Sect: 4   Obj: 2   RANDOM: Y

66. Construct a frequency polygon: The following data revealed the number of hours of television watched by 130 five-year-olds on a Saturday.

Class Limits	Frequency
0–2	15
3–5	40
6–8	45
9–11	20
12–14	10

Answer:

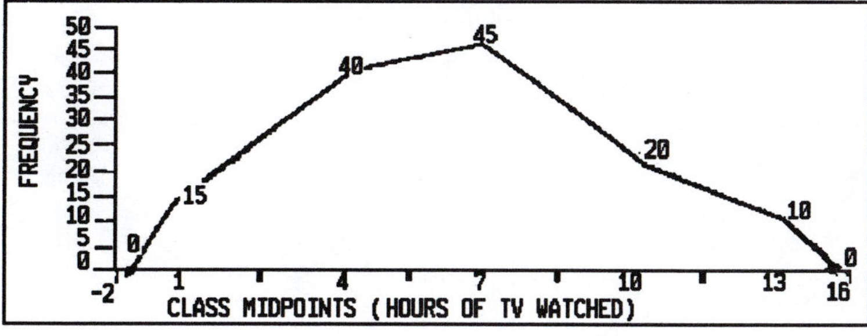

Type: S   Sect: 4   Obj: 2   RANDOM: Y

67. Construct a frequency polygon: The tourism department in Aspen, Colorado, wanted to know the length of stay at their resorts. They collected the following data:

Length of Stay (days)	Frequency
3	8
4	10
5	15
6	20
7	28
8	18
9	12
10	4

Answer:

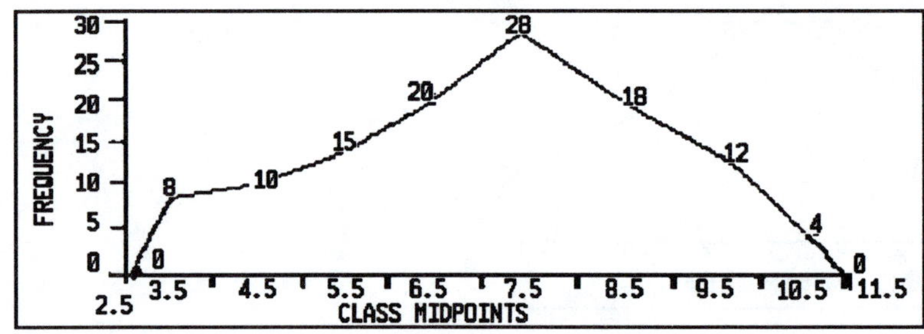

Type: S   Sect: 4   Obj: 2   RANDOM: Y

68. Construct a frequency polygon: The following data indicate the monthly salaries of a sample of 70 nurses.

Salary	Frequency
1,900	5
2,200	10
2,500	20
2,800	25
3,100	8
3,400	2

Answer:

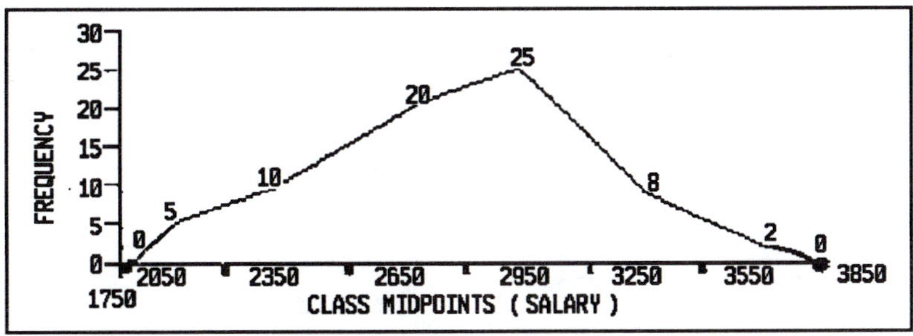

Type: S   Sect: 4   Obj: 2   RANDOM: Y

69. Construct a frequency polygon: The following is the distribution of the number of mistakes 70 students made when translating an article from Spanish to English.

Class Limits	Frequency
0–4	18
5–9	29
10–14	13
15–19	8
20–24	2

Answer:

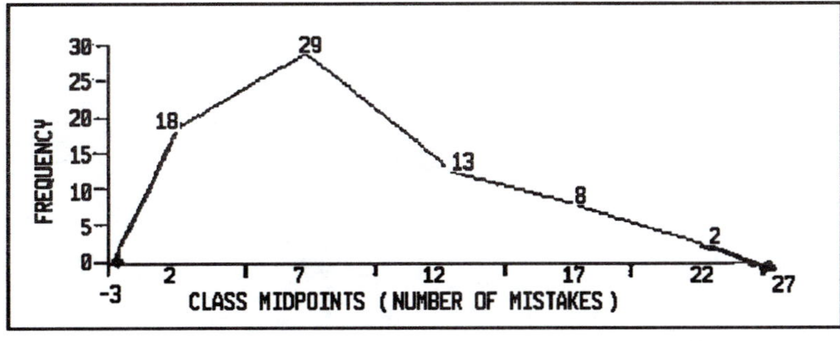

Type: S   Sect: 4   Obj: 2   RANDOM: Y

70. Construct an ogive of number of cups "or less": The number of cups of coffee drunk by 150 employees at Sears were as follows:

No. of Cups	Frequency
0	15
1	20
2	30
3	45
4	25
5	15

Answer:

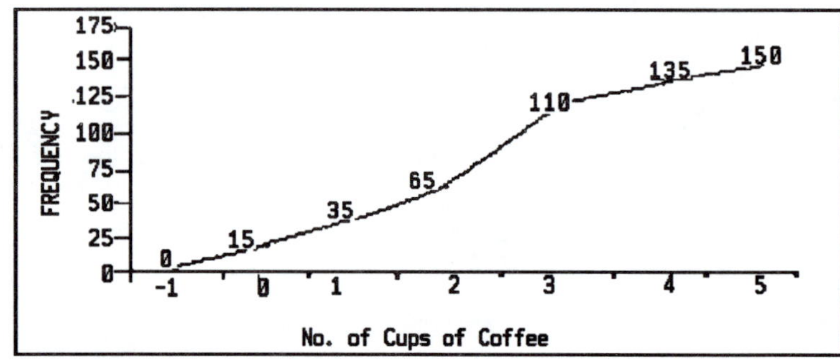

Type: S   Sect: 6 Obj: 2   RANDOM: Y

71. Construct an ogive of the number of hours of TV "or less": The following data revealed the number of hours of television watched by 130 five-year-olds on a Saturday.

Class Limits	Frequency
0–2	15
3–5	40
6–8	45
9–11	20
12–14	10

Answer:

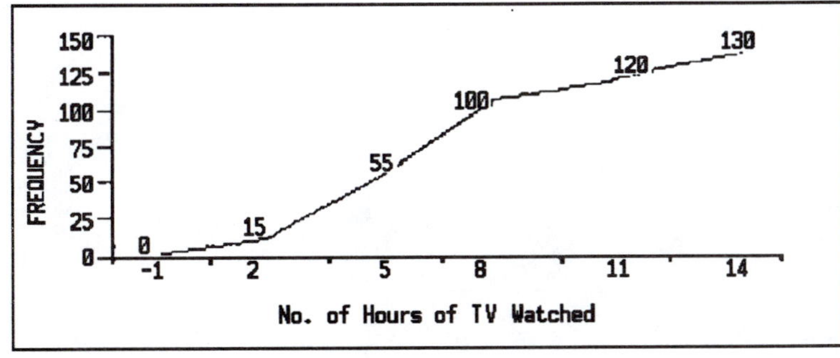

Type: S   Sect: 6 Obj: 2   RANDOM: Y

72. Construct an ogive of length of stay "or less": The tourism department in Aspen, Colorado, wanted to know the length of stay at their resorts. They collected the following data:

Length of Stay (days)	Frequency
3	8
4	10
5	15
6	20
7	28
8	18

Answer:

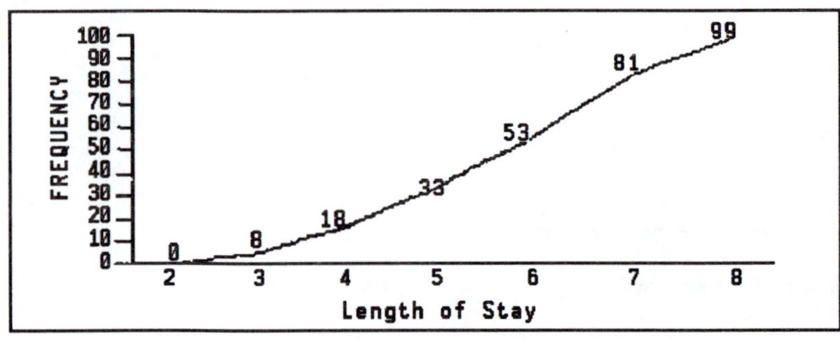

Type: S   Sect: 6   Obj: 2   RANDOM: Y

73.Construct an ogive of monthly salary "or less": The following data indicate the monthly salaries of a sample of 70 nurses.

<u>Salary</u>	<u>Frequency</u>
1,900	5
2,200	10
2,500	20
2,800	25
3,100	8
3,400	2

Answer:

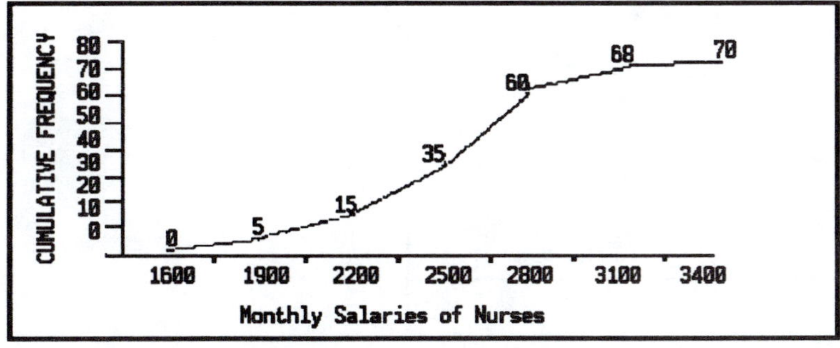

Type: S   Sect: 6 Obj: 2   RANDOM: Y

74. Construct an ogive of days missed "or less": The absentee records over the past year for employees of Hugo's Dairy are listed below.

Days Missed	Frequency
0	9
1	12
2	8
3	5
4	10
5	5

Answer:

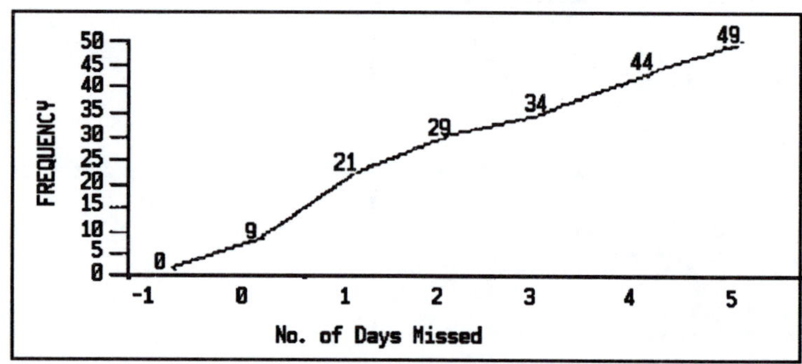

Type: S   Sect: 6   Obj: 2   RANDOM: Y

75. Construct an ogive of number of mistakes "or less": The following is the distribution of the number of mistakes 70 students made when translating an article from Spanish to English.

Class Limits	Frequency
0–4	18
5–9	29
10–14	13
15–19	8
20–24	2

Answer:

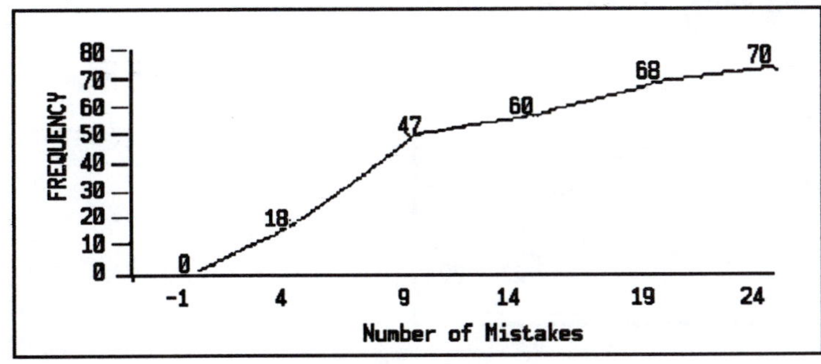

Type: S   Sect: 6 Obj: 2   RANDOM: Y

76. Construct a frequency polygon and an ogive for number of years "or less" for the
following data: For 85 employees of a large corporation, the following distribution for
years of service was obtained.

Class Limits	Frequency
1–5	20
6–10	27
11–15	15
16–20	13
21–25	10

Answer:

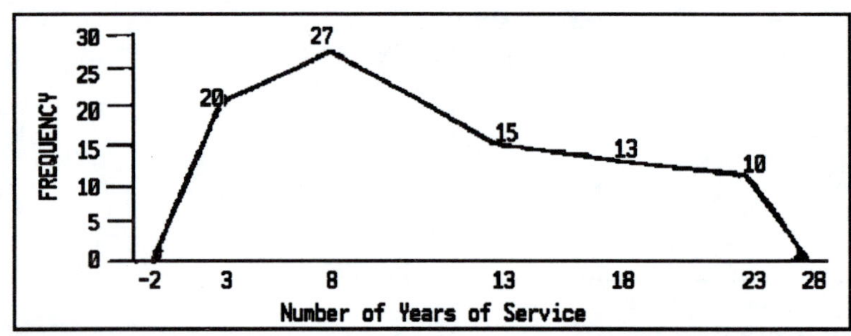

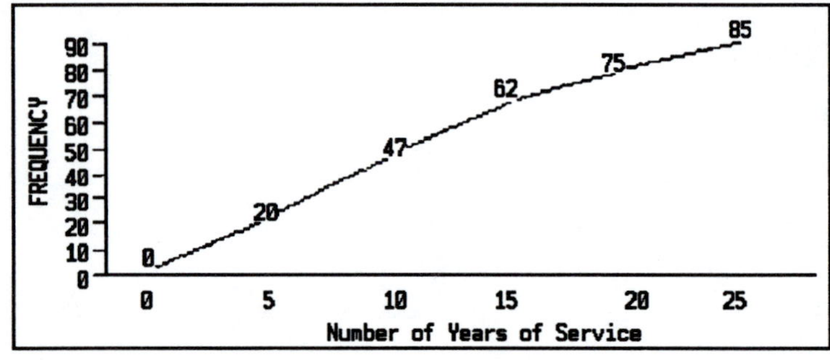

Type: S   Sect: 4 Obj: 2   RANDOM: Y

77. Construct a frequency polyogon and an ogive for amount of cholesterol "or less" for the following data: The cholesterol levels of selected patients are listed below.

Class Limits	Frequency
180–184	15
185–189	18
190–194	9
195–199	12
200–204	10
205–209	15

Answer:

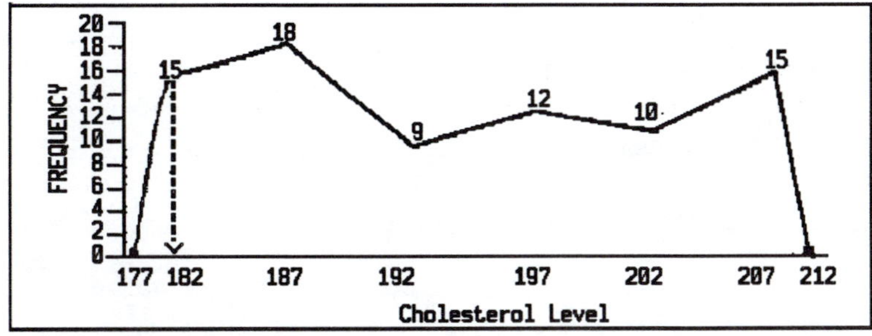

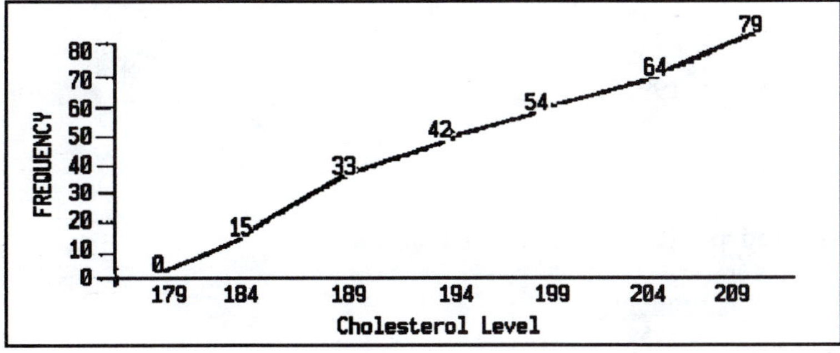

Type: S   Sect: 4   Obj: 2   RANDOM: Y

78. The cholesterol levels of selected patients are listed below.

Class Limits	Frequency
180–184	15
185–189	18
190–194	9
195–199	12
200–204	10
205–209	15
210–214	11

How many values are in the class 190–194?

Answer: 9 Type: S   Sect: 2 Obj: 2   RANDOM: Y

79. The cholesterol levels of selected patients are listed below.

Class Limits	Frequency
180–184	15
185–189	18
190–194	9
195–199	12
200–204	10
205–209	15
210–214	11

How many values are between 184.5 and 199.5?

Answer: 39 Type: S   Sect: 3 Obj: 2   RANDOM: Y

80. The cholesterol levels of selected patients are listed below.

Class Limits	Frequency
180–184	15
185–189	18
190–194	9
195–199	12
200–204	10
205–209	15
210–214	11

How many values are below 210?

Answer: 79 Type: S   Sect: 3 Obj: 2   RANDOM: Y

81. The cholesterol levels of selected patients are listed below.

Class Limits	Frequency
180–184	15
185–189	18
190–194	9
195–199	12
200–204	10
205–209	15
210–214	11

How many values are above 194?

Answer: 48   Type: S   Sect: 3   Obj: 2   RANDOM: Y

82. The following data revealed the number of hours of television watched by 130 five-year-olds on a Saturday.

Class Limits	Frequency
0–2	15
3–5	40
6–8	45
9–11	20
12–14	10

How many values are above 11?

Answer: 10   Type: S   Sect: 3   Obj: 2   RANDOM: Y

83. The following data revealed the number of hours of television watched by 130 five-year-olds on a Saturday.

Class Limits	Frequency
0–2	15
3–5	40
6–8	45
9–11	20
12–14	10

How many values are between 2.5 and 8.5?

Answer: 85   Type: S   Sect: 3   Obj: 2   RANDOM: Y

84. The following data revealed the number of hours of television watched by 130 five-year-olds on a Saturday.

Class Limits	Frequency
0–2	15
3–5	40
6–8	45
9–11	20
12–14	10

How many values are less than 9?

Answer: 100 Type: S   Sect: 3 Obj: 2   RANDOM: Y

85. The numbers of music lessons given each day for 20 days are listed below. Construct a stem-and-leaf display for the data.

```
15   10    8   18   25   24    7   19   22   32
11   15   22    6   13   19   21   16    9   11
```

Answer:
```
0 | 6  7  8  9
1 | 0  1  1  3  5  5  6  8  9  9
2 | 1  2  2  4  5
3 | 2
```
Type: S   Sect: 7 Obj: 2   RANDOM: Y

86. A sample of 25 boxes of cereal gives the following weights. Construct a stem-and-leaf display.

```
15.2     15.1     14.8     13.2     14.6
14.6     14.7     15.0     13.9     15.2
15.4     14.8     13.8     14.8     15.1
15.0     15.1     13.8     14.9     13.8
15.1     15.2     14.7     14.8     14.9
```

Answer:
```
13 | 2  8  8  8  9
14 | 6  6  7  7  8  8  8  8  9  9
15 | 0  0  1  1  1  1  2  2  2  4
```
Type: S   Sect: 7 Obj: 2   RANDOM: Y

87.A stem-and-leaf display was constructed as follows:

```
1 | 1  2  2
2 | 3  4  4  5  6
3 | 0  2  3  3  4  5
4 | 0  1  1  2
5 | 1  2  2  5  6  6  7
```

How many items were in the sample?

Answer: 25 Type: S   Sect: 7 Obj: 2   RANDOM: Y

MULTIPLE CHOICE

****
The remaining questions in this chapter require the use of the formula
generation capabilities of ESATEST III. Please see the ESATEST III
documentation if you need directions on how to use them.
****

88.For the data @5, 9,@2, 11, 7, @19, and@8, what is the range?
   a. l4
   b. l3
   c. l14
   d. l17

   Answer: @3-@2,5,5,10,0,2,1,3,0,19,15,25,0,8,5,9,0 Type: M   RANDOM: F

SHORT ANSWER

89.Find the class boundaries. @8-@18

   Answer: `~1-0.5` –̇ `~2+0.5` ,8,5,10,0,18,15,20,0 Type: S   RANDOM: F

90.Find the class midpoint. @7-@11

   Answer: (@1+@2)/2,7,1,9,(1,3,5,7,9),11,11,19,0,(11,13,15,17,19) Type: S   RANDOM: F

91.Determine the class size. @18-@22

   Answer: @2-@1+1,18,15,20,0,22,21,25,0 Type: S   RANDOM: F

CHAPTER 3

TRUE/FALSE

1. The measure that balances the distances of the values from the center is called the mode.

   Answer: F Type: T   Sect: 1 Obj: 1   RANDOM: Y

2. Another name for the arithmetic average is the mean.

   Answer: T Type: T   Sect: 1 Obj: 1   RANDOM: Y

3. For the sample 1, 8, 7, 2, 9, 15, and 18, the mean is 7.6.

   Answer: F Type: T   Sect: 1 Obj: 1   RANDOM: Y

4. For the sample 18, 52, 19, 17, 18, 19, 19, and 21, the mode is 19.

   Answer: T Type: T   Sect: 1 Obj: 1   RANDOM: Y

5. For the sample 1.3, 2.7, 8.9, 7.3, 9.2, and 8.1, the median is 7.7.

   Answer: T Type: T   Sect: 1 Obj: 1   RANDOM: Y

6. For the data 2.8, 1.9, 2.7, 8.9, and 9.7, the mean is 3.9.

   Answer: F Type: T   Sect: 1 Obj: 1   RANDOM: Y

7. For the data 15, 18, 12, 17, 13, 16, and 20, the median is 17.

   Answer: F Type: T   Sect: 1 Obj: 1   RANDOM: Y

8. The data set 15, 12, 18, 17, 2, and 19 has a range of 4.

   Answer: F Type: T   Sect: 7 Obj: 2   RANDOM: Y

9. If the values of a data set are near the mean, the variance will be small.

   Answer: T Type: T   Sect: 7 Obj: 2   RANDOM: Y

10. The standard deviation of the data 8, 12, 15, 7, 11, and 10 is 2.88.

    Answer: T Type: T   Sect: 8 Obj: 2   RANDOM: Y

11. The variance of the data 19, 25, 17, 21, 18, 15, and 24 is 10.5.

    Answer: F Type: T   Sect: 8 Obj: 2   RANDOM: Y

12. The mode of the data set 5, 7, 2, 1, 5, 9, and 3 is 5.

    Answer: T Type: T   Sect: 1 Obj: 1   RANDOM: Y

13. Jerry got a score of 92 on his history test. The test had a mean of 80 and a standard deviation of 11. The z-score for Jerry was 1.09.

    Answer: T Type: T   Sect: 8 Obj: 2   RANDOM: Y

14. Mrs. Oliver gave a quiz to her students. Their scores were 8, 7, 9, 10, 7, 8, 6, 4, 5, and 9. The mean of these scores was 8.

    Answer: F Type: T   Sect: 1 Obj: 1   RANDOM: Y

15. For the data 10, 11, 15, 16, 12, 19, 18, 17, 19, 20, 13, and 20, the value corresponding to the 63rd percentile would be 17.

    Answer: F Type: T   Sect: 4 Obj: 3   RANDOM: Y

MULTIPLE CHOICE

16. If you look at the coins in your pocket and state that you have more dimes than any other coin, what measure of central tendency are you using?
    a. mode
    b. mean
    c. midrange
    d. median

    Answer: a Type: M   Sect: 1 Obj: 1   RANDOM: Y

17. What is the midpoint of the data set called?
    a. mode
    b. median
    c. mean
    d. midrange

    Answer: b Type: M   Sect: 1 Obj: 1   RANDOM: Y

18.For the data 15, 21, 28, 17, 19, 29, and 25, what is the mean?
  a. 21
  b. 19
  c. 20
  d. 22

  Answer: d Type: M   Sect: 1 Obj: 1   RANDOM: Y

19.What is the median of the data set 120, 117, 191, 115, 130, 125, 140, and 126?
  a. 125
  b. 126
  c. 125.5
  d. 122.5

  Answer: c Type: M   Sect: 1 Obj: 1   RANDOM: Y

20.The following data set was collected: 32, 37, 38, 36, 37, 39, 37, 35, 41, and 31. What
  is the mode?
  a. 37
  b. 36
  c. 35
  d. 38

  Answer: a Type: M   Sect: 1 Obj: 1   RANDOM: Y

21.The following data were reported on weekly salaries:

Salary	Frequency
300	4
350	5
400	8
450	10
500	9
550	12

What is the mean of this distribution?
  a. 453.15
  b. 362.50
  c. 425.00
  d. 435.75

  Answer: a Type: M   Sect: 3 Obj: 1   RANDOM: Y

22. The following data were reported on weekly salaries:

Salary	Frequency
300	4
350	5
400	8
450	10
500	9
550	12

What is the mode?
a. 12
b. 550
c. 450
d. 500

Answer: b Type: M   Sect: 2 Obj: 1   RANDOM: Y

23. The following frequency distribution was obtained:

Days	Frequency
17–23	8
24–30	3
31–37	9
38–44	7
45–51	3

What is the mean?
a. 38.50
b. 32.6
c. 33.6
d. 30.5

Answer: b Type: M   Sect: 3 Obj: 1   RANDOM: Y

24. What is the weighted mean of the cost of tennis shoes?

Shoe	Number Purchased	Price
1	8	$35
2	10	$45
3	14	$25
4	6	$55

a. $40
b. $37.11
c. $38.27
d. $39.27

Answer: b Type: M   Sect: 3 Obj: 1   RANDOM: Y

25. The following data were collected: 38, 42, 37, 50, 32, 55. What is the variance?
    a. 42.33
    b. 8.64
    c. 62.22
    d. 74.67

Answer: d Type: M   Sect: 8 Obj: 2   RANDOM: Y

26. The mean of a sample was 375 and the variance was 88. What is the standard deviation?
    a. 7744
    b. 9.4
    c. 19.4
    d. 4.3

Answer: b Type: M   Sect: 8 Obj: 2   RANDOM: Y

27. Charlie had a z-score of −1.54 on a test that had a mean of 72 and a standard deviation
    of 13. What was Charlie's raw score on the test?
    a. 52
    b. 92
    c. 70
    d. 54

Answer: a Type: M   Sect: 8 Obj: 2   RANDOM: Y

28. A sample has a mean of 115 and a standard deviation of 22. According to Chebyshev's
    theorem, at least 75% of the data will fall between what values?
    a. 49 and 181
    b. 93 and 137
    c. 71 and 159
    d. 105 and 125

Answer: c Type: M   Sect: 8 Obj: 2   RANDOM: Y

29. A sample has a mean of 75 and a standard deviation of 10. According to the empirical
    rule, about 68% of the data will fall between what values?
    a. 70 and 80   b. 65 and 85
    c. 55 and 95   d. 45 and 105

Answer: b Type: M   Sect: 8 Obj: 2   RANDOM: Y

30.For the data set 17, 25, 30, 27, 16, 19, 35, 42, 18, 20, 19, and 13, what value
corresponds to the 31st percentile?
a. 18.5
b. 27
c. 17
d. 21.5

Answer: a Type: M   Sect: 4 Obj: 3   RANDOM: Y

SHORT ANSWER

Directions for the problems below: Find the first quartile of the given data.

31.Nine companies reported the following numbers of telephone calls asking for information
each day: 38, 52, 40, 40, 60, 65, 42, 58, and 61.

Answer: 40 Type: S   Sect: 4 Obj: 3   RANDOM: Y

32.For 12 years, a meteorologist kept track of the number of days it snowed in February.
The results were as follows:

3, 5, 7, 2, 8, 10, 6, 5, 7, 15, 2, 7

Answer: 4 Type: S   Sect: 4 Obj: 3   RANDOM: Y

33.Ten dieters kept track of how many glasses of water they drank each day.   Their data
are: 4, 6, 7, 6, 5, 6, 4, 8, 6, 7.

Answer: 5.5 Type: S   Sect: 4 Obj: 3   RANDOM: Y

34.The weights (in pounds) of 15 models were recorded as follows: 105, 110, 120, 118, 105,
108, 110, 117, 125, 105, 113, 105, 119, 121, and 120.

Answer: 105 Type: S   Sect: 4 Obj: 3   RANDOM: Y

35.The price of unleaded gasoline was reported at 9 different stations around town.   The
results were: $1.19, 1.25, 1.26, 1.20, 1.25, 1.29, 1.21, 1.25, and 1.24.   Find the
third quartile.

Answer: 1.255 Type: S   Sect: 4 Obj: 3   RANDOM: Y

Directions for the problems below: Find the mean, median, and mode.

36. A bowler recorded the following bowling scores over the last 10 games: 137, 162, 203, 170, 165, 162, 143, 194, 159, 173.

Answer: mean = 166.8, median = 163.5, mode = 162   Type: S   Sect: 1   Obj: 1

RANDOM: Y

37. A class of 15 second-grade students was given a test to see how many simple addition problems they could do in 3 minutes.   The results were as follows: 15, 25, 32, 18, 23, 35, 16, 23, 30, 19, 27, 33, 17, 29, 23.

Answer: mean = 24.3, median = 23, mode = 23   Type: S   Sect: 1   Obj: 1   RANDOM: Y

38. The rainfall in inches for 8 months in a portion of Florida was recorded as: 2.8, 3.6, 6.6, 7.6, 6.2, 7.6, 9.5, 8.2.

Answer: mean = 6.51, median = 7.1, mode = 7.6   Type: S   Sect: 1   Obj: 1   RANDOM: Y

39. The following data represent the number of points scored by the Tigers in their last 11 football games: 20, 18, 30, 10, 21, 6, 18, 35, 12, 15, 27.

Answer: mean = 19.3, median = 18, mode = 18   Type: S   Sect: 1   Obj: 1   RANDOM: Y

40. The following data represent the number of morning papers in circulation in 10 different states: 15, 13, 10, 5, 11, 4, 6, 11, 20, and 7.

Answer: mean = 10.2, median = 10.5, mode = 11   Type: S   Sect: 1   Obj: 1   RANDOM: Y

41. The following distribution of the number of cars owned by each of 450 families is listed below.

No. of Cars Owned	Number of Families
0	20
1	125
2	270
3	30
4	5

What is the mean of this distribution?

Answer: 1.7   Type: S   Sect: 3   Obj: 1   RANDOM: Y

42. The following frequency distribution shows the number of students who received the given points on a 10-point quiz.

Points on Quiz	No. of Students
4	2
5	5
6	7
7	12
8	20
9	8
10	3

What is the mean of this distribution?

Answer: 7.4 Type: S   Sect: 3 Obj: 1   RANDOM: Y

43. The weekly salaries of 140 people at Toyo are listed below.

Salaries	f
175-199	5
200-224	10
225-249	18
250-274	25
275-299	45
300-324	22
325-349	10
350-374	5

What is the mean of this distribution?

Answer: 277.4 Type: S   Sect: 3 Obj: 1   RANDOM: Y

44. The following frequency distribution lists the ages of associate professors at St. James College.

Ages	Frequency
26-34	3
35-43	10
44-52	30
53-61	55
62-70	7

What is the mean of this frequency distribution?

Answer: 52.5 Type: S   Sect: 3 Obj: 1   RANDOM: Y

45. The following frequency distribution lists the number of hours spent by 900 people at a popular amusement park.

Number of Hours	Frequency
0-4	20
5-9	150
10-14	375
15-19	100
20-24	65
25-29	150
30-34	25
35-39	15

What is the mean of this frequency distribution?

Answer: 15.7 Type: S   Sect: 3 Obj: 1   RANDOM: Y

46. A college professor assigns grades based on the following breakdown: homework--10% of the grade, chapter tests--70%, and the final exam--20%. A student had a homework average of 88, a test average of 82, and a final exam score of 79. Find the final grade, using the weighted mean.

Answer: 82 Type: S   Sect: 3 Obj: 1   RANDOM: Y

47. A temporary service employs the following types of workers, along with their hourly wages:

8 File clerks	$6.50
35 Secretaries	$8.00
10 Receptionists	$6.25
9 Stenographers	$8.50

Find the weighted mean hourly salary.

Answer: $7.60 Type: S   Sect: 3 Obj: 1   RANDOM: Y

48. Find the weighted mean price of 3 different kinds of trucks. The number and price of each are shown below.

	Number	Price
Model A	60	$15,600
Model B	100	$ 9,800
Model C	40	$11,700

Answer: $11,920 Type: S   Sect: 3 Obj: 1   RANDOM: Y

49. Find the weighted mean of the numbers of hours 40 students spent studying for their calculus final.

No. of Students	No. of Hours
6	4
15	6
14	8
5	10

Answer: 6.9 Type: S   Sect: 3 Obj: 1   RANDOM: Y

50. A stockbroker calculates the following percentages of each of four stock investments paying off as follows:

	Percentage	Payoff
Stock A	15%	6000
Stock B	25%	10000
Stock C	32%	3000
Stock D	28%	5000

What is the average payoff? Use the weighted mean.

Answer: $5760 Type: S   Sect: 3 Obj: 1   RANDOM: Y

51. What is the symbol used to represent the standard deviation of a population?

Answer: σ Type: S   Sect: 8 Obj: 2   RANDOM: Y

52. For the numbers 6, 8, 15, 25, and 30, what is the mean?

Answer: 16.8 Type: S   Sect: 1 Obj: 1   RANDOM: Y

53. The weekly amounts spent on food by 110 families of three are listed below.

Class Limits	Frequency
57-65	6
66-74	12
75-83	23
84-92	19
93-101	25
102-110	17
111-119	8

Answer: variance = 213.32, standard deviation = 14.61 Type: S   Sect: 11 Obj: 2

RANDOM: Y

54.The following table shows the number of defective parts in 75 different shipments.

No. of Defective Parts	Frequency
0	20
1	25
2	15
3	10
4	5

Answer: variance = 1.05, standard deviation = 1.03 Type: S   Sect: 11 Obj: 2

RANDOM: Y

55.The following is the distribution of the numbers of mistakes 70 students made translating an article from Spanish to English.

Class Limits	Frequency
0-4	18
5-9	29
10-14	13
15-19	8
20-24	2

Answer: variance = 27.85, standard deviation = 5.28 Type: S   Sect: 11 Obj: 2

RANDOM: Y

56.A survey of 50 selected customers showed the numbers of days it took to receive their new magazine subscriptions.

Class Limits	Frequency
20-30	4
31-41	9
42-52	14
53-63	20
64-74	3

Answer: variance = 136.75, standard deviation = 11.69 Type: S   Sect: 11 Obj: 2

RANDOM: Y

57. The following distribution shows the numbers of ounces of 55 randomly selected ribeye steaks at the supermarket.

Class Boundaries	Frequency
5.6-6.4	7
6.5-7.3	10
7.4-8.2	18
8.3-9.1	13
9.2-10.0	7

Answer: variance = 1.183, standard deviation = 1.087 Type: S   Sect: 11 Obj: 2

RANDOM: Y

58. The average age of a resident of Honolulu was thought to be 77 with a standard deviation of 4.5 years. What is the z-score of an 85-year-old woman?

Answer: 1.78 Type: S   Sect: 8 Obj: 2   RANDOM: Y

59. The average hourly salary of plumbers in Des Moines was thought to be $37 with a standard deviation of $3. What is the z-score for a plumber who makes $48?

Answer: 3.67 Type: S   Sect: 8 Obj: 2   RANDOM: Y

60. The average hourly salary of plumbers in Des Moines was thought to be $37 with a standard deviation of $3. What is the z-score for a plumber who makes $30?

Answer: -2.33 Type: S   Sect: 8 Obj: 2   RANDOM: Y

61. The average hourly salary of plumbers in Des Moines was thought to be $37 with a standard deviation of $3. What is the z-score for a plumber who makes $43?

Answer: 2 Type: S   Sect: 8 Obj: 2   RANDOM: Y

62. At Ramey's supermarket, the average waiting time at the checkout counter is 5 minutes with a standard deviation of 1.5 minutes. If a woman stands in line for 8 minutes, what is her z-score?

Answer: 2 Type: S   Sect: 8 Obj: 2   RANDOM: Y

63. At Ramey's supermarket, the average waiting time at the checkout counter is 5 minutes with a standard deviation of 1.5 minutes. If a man stands in line for 4 minutes, what is his z-score?

Answer: -0.67 Type: S   Sect: 8 Obj: 2   RANDOM: Y

64. The Warrensville basketball team scores an average of 75 points per game with a standard deviation of 8 points. If the team scored 90 points in its next game, what was its z-score?

Answer: 1.88 Type: S   Sect: 8 Obj: 2   RANDOM: Y

65. The average number of calories in a piece of pie was 250 with a standard deviation of 30.   If a randomly selected piece of pie has a z-score of -1.23, how many calories does it contain?

Answer: approximately 213 calories Type: S   Sect: 8 Obj: 2   RANDOM: Y

66. Sandy and Dennis want to compare their scores on an English exam. Sandy got a grade of 80 on a test with a mean of 70 and a standard deviation of 8. Dennis got a 72 on the test with a mean of 65 and a standard deviation of 9. Which of the students had the better relative score and what was that score?

Answer: Sandy had a better relative score. Her z-score was 1.25 and Dennis's z-score was 0.78.

Type: S   Sect: 8 Obj: 2   RANDOM: Y

67. The average score of students in Mr. Blake's science class was 73 with a standard deviation of 11, while the average score of students in Mrs. Arnold's class was 75 with a standard deviation of 10. Which class is more variable?

Answer: Mr. Blake's class is more variable with a coefficient of variation of 0.151.

Type: S   Sect: 2 Obj: 2   RANDOM: Y

68. The average price of a Honda was \$13,500 with a standard deviation of \$700. The average price of a Nissan was \$12,500 with a standard deviation of \$625. What are the coefficients of variation of both models, and which is more variable?

Answer:
The Honda's price is more variable, with a coefficient of variation of 0.052, while the price of the Nissan has a coefficient of variation of 0.05.
Type: S   Sect: 2 Obj: 2   RANDOM: Y

69. The average hourly wage at Burger Chef was \$4.30 with a standard deviation of 0.25, while the average hourly wage at Burger Haven was \$3.95 with a standard deviation of 0.30. What are the variations of wages at both places, and which is more variable?

Answer:
The wage at Burger Haven is more variable, with a coefficient of variation of 0.076, while the coefficient of variation of Burger Chef's wage is 0.058.
Type: S   Sect: 2 Obj: 2   RANDOM: Y

70. The mean of a distribution is 82 with a standard deviation of 14. According to Chebyshev's theorem, at least what percent of the data will fall between 54 and 110?

    Answer: 75% Type: S   Sect: 8 Obj: 2   RANDOM: Y

71. The mean of a distribution is 35 and the standard deviation is 4. At least what percent of the data will fall between 23 and 47?

    Answer: 89% Type: S   Sect: 8 Obj: 2   RANDOM: Y

72. What is the symbol used to represent the mean of a sample?

    Answer:

    $\overline{X}$

    Type: S   Sect: 1 Obj: 1   RANDOM: Y

73. What is the symbol used to represent the mean of a population?

    Answer: μ Type: S   Sect: 1 Obj: 1   RANDOM: Y

74. In a set of sample data, 8 scores are smaller than 65 and 8 scores are larger than 65. We would say that 65 is what measure of central tendency?

    Answer: median Type: S   Sect: 1 Obj: 1   RANDOM: Y

75. A shoe salesman reports that more women buy size 7 shoes than any other size. What type of measure is being used?

    Answer: mode Type: S   Sect: 2 Obj: 1   RANDOM: Y

76. For the data set 5, 16, 7, 8, 11, and 19, what is the value of the third quartile?

    Answer: 17 Type: S   Sect: 4 Obj: 3   RANDOM: Y

77. For the data set 8, 12, 15, 7, 19, 16, 20, 36, and 17, what is the value of the third quartile?

    Answer: 19.5 Type: S   Sect: 4 Obj: 3   RANDOM: Y

78. For the data set 25, 16, 32, 40, 18, 50, 19, and 60, what is the value of the third quartile?

    Answer: 45 Type: S   Sect: 4 Obj: 3   RANDOM: Y

79.For the data set 100, 85, 93, 92, 76, 82, and 65, what is the value of the third
quartile?

   Answer: 93 Type: S   Sect: 4 Obj: 3   RANDOM: Y

80.For the data set 100, 85, 93, 92, 76, 82, and 65, what is the value of the first
quartile?

   Answer: 76 Type: S   Sect: 4 Obj: 3   RANDOM: Y

81.Suppose you got a 75 on your math test, and your teacher told you that approximately 23%
of the scores were higher than yours. At what percentile does your score place you?

   Answer: 77th percentile Type: S   Sect: 4 Obj: 3   RANDOM: Y

82.Suppose you get a 125 on your Statistics final and your teacher told you that
approximately 35% of the scores were higher than yours. At what percentile does your
score place you?

   Answer: 65th percentile Type: S   Sect: 4 Obj: 3   RANDOM: Y

83.If you take a placement test and approximately 78% of the scores are lower than yours,
at what percentile are you located?

   Answer: 78th percentile Type: S   Sect: 4 Obj: 3   RANDOM: Y

84.For the data set 20, 13, 18, 7, 22, 14, 18, 25, 9, 11, 15, 25, 16, and 35, what
percentile is the number 22?

   Answer: about the 71st percentile Type: S   Sect: 4 Obj: 3   RANDOM: Y

85.For the data set 13, 20, 7, 18, 14, 22, 18, 9, 11, 25, 15, 16, 25, and 35, what
percentile is the number 13?

   Answer: the 21st percentile Type: S   Sect: 4 Obj: 3   RANDOM: Y

86.For the data set 2, 7, 9, 15, 18, 20, 6, 9, 15, 11, 25, 23, 18, 16, 3, 6, 18, 19, and
24, what percentile is the number 7?

   Answer: the 21st percentile Type: S   Sect: 4 Obj: 3   RANDOM: Y

87.For the data set 2, 7, 9, 15, 18, 20, 6, 9, 15, 11, 25, 23, 18, 16, 3, 6, 18, 19, and
24, what percentile is the number 19?

   Answer: approximately the 74th percentile Type: S   Sect: 4 Obj: 3   RANDOM: Y

88. The number of hours of television watched per week by 25 sixth graders were collected. The eight smallest values in the data set were 5, 12, 10, 7, 6, 11, 14, and 15. Using this information, what is the 23rd percentile?

    Answer: 11.5 Type: S   Sect: 4 Obj: 3   RANDOM: Y

89. If the first quartile is 101.5 and the third quartile is 150, what is the interquartile range?

    Answer: 48.5 Type: S   Sect: 7 Obj: 2   RANDOM: Y

90. If the first quartile is 60.5 and the third quartile is 87.5, what is the interquartile range?

    Answer: 27 Type: S   Sect: 7 Obj: 2   RANDOM: Y

91. If the first quartile is 55 and the third quartile is 76, what is the interquartile range?

    Answer: 21 Type: S   Sect: 7 Obj: 2   RANDOM: Y

92. For the data set 5, 12, 16, 22, 11, and 18, what is the mean absolute deviation?

    Answer: 4.67 Type: S   Sect: 7 Obj: 2   RANDOM: Y

93. For the data set 25, 16, 19, 12, 8, 24, 15, 18, 21, and 17, what is the mean absolute deviation?

    Answer: 3.9 Type: S   Sect: 7 Obj: 2   RANDOM: Y

94. For the data set 15, 21, 28, 17, 19, 29, and 25, what is the mean absolute deviation?

    Answer: 4.6 Type: S   Sect: 7 Obj: 2   RANDOM: Y

95. For the data set 58, 63, 75, 82, 95, 98, 75, 60, 87, 72, 83, and 82, what is the mean absolute deviation?

    Answer: 10.3 Type: S   Sect: 7 Obj: 2   RANDOM: Y

96. A distribution has a mean of 135 and a standard deviation of 12. According to Chebyshev's rule, at least 89% of the values would fall within how many standard deviations?

    Answer: 3 Type: S   Sect: 8 Obj: 2   RANDOM: Y

97.A distribution has a mean of 500 and a standard deviation of 30. Chebyshev's rule says
that at least 75% of the data will fall between what values?

Answer: 440 and 560 Type: S   Sect: 8 Obj: 2   RANDOM: Y

98.A bell-shaped distribution has a mean of 460 and a standard deviation of 16. According
to the empirical rule, we would expect about 95% of the data to fall between what
values?

Answer: 428 and 492 Type: S   Sect: 8 Obj: 2   RANDOM: Y

99.For the data set 15, 21, 19, 32, 11, 18, and 30, what is the value of the median?

Answer: 19 Type: S   Sect: 1 Obj: 3   RANDOM: Y

100.What is an extremely high or low value in a data set called?

Answer: an outlier Type: S   Sect: 2 Obj: 3   RANDOM: Y

101.The value of the first quartile for a set of data is 30 and the value of the third
quartile is 78. What is the interquartile range?

Answer: 48 Type: S   Sect: 7 Obj: 2   RANDOM: Y

MULTIPLE CHOICE

****
The remaining questions in this chapter require the use of the formula
generation capabilities of ESATEST III. Please see the ESATEST III
documentation if you need directions on how to use them.
****

102.For the data 15, 21, @28, 17, @19, 29, and @25, what is the mean?
   a. | 21.12
   b. | 19.52
   c. | 20.96
   d. | 22.39

Answer: (82+@1+@2+@3)/7,28,25,30,0,19,15,20,0,25,20,30,0,[2]   Type: M   RANDOM: F

103.What is the median of the data set @120, 117, @191, @115, 130, 125, 140, and 126?
   a. I 125
   b. I 126
   c. I 125.5
   d. I 122.5

Answer: @1-@1+125.5,120,115,120,0,191,180,200,0,115,100,120,0 Type: M   RANDOM: F

104.The following data set was collected: @32, 37, @38, 36, 37, 39, 37, 35, @41, and 31.
   What is the mode?
   a. I 37
   b. I 36
   c. I 35
   d. I 38

Answer: @1-@1+37,32,30,34,0,38,35,40,0,41,40,45,0 Type: M   RANDOM: F

105.The following data were collected: 38, 42, 37, 50, @32, `87-~1`. What is the variance?
   a. I 42.33
   b. I 8.64
   c. I 62.22
   d. I 74.67

Answer: (@1^2+(87-@1)^2-3675.666667)/5,32,20,40,0,[2]   Type: M   RANDOM: F

106.The mean of a sample was 375 and the variance was `@88`. What is the standard
   deviation?
   a. I 7744
   b. I 9.4
   c. I 19.4
   d. I 4.3

Answer: @1^0.5,88,70,90,0,[1]   Type: M   RANDOM: F

107.For the data set 17, 25, 30, @27, 16, 19, 35, @42, 18, 20, 19, and @13, what value
   corresponds to the 31st percentile?
   a. I 18.5
   b. I 27
   c. I 17
   d. I 21.5

Answer: @1-@1+18.5,27,20,30,0,42,40,60,0,13,5,15,0 Type: M   RANDOM: F

CHAPTER 4

TRUE/FALSE

1. A compound event consists of two or more outcomes or simple events.

   Answer: T Type: T Sect: 7 Obj: 1 RANDOM: Y

2. James got 5 ones in 12 rolls of a die. This is an example of subjective probability.

   Answer: F Type: T Sect: 2 Obj: 1 RANDOM: Y

3. If a coin is tossed three times, the probability that all three tosses are tails is 1/8.

   Answer: T Type: T Sect: 2 Obj: 2 RANDOM: Y

4. A forecaster says there is a 30% chance of freezing rain today. This is an example of subjective probability.

   Answer: T Type: T Sect: 2 Obj: 1 RANDOM: Y

5. A card is drawn from an ordinary deck. The probability that it is a diamond is 1/3.

   Answer: F Type: T Sect: 2 Obj: 2 RANDOM: Y

6. A single die is rolled. The events "getting a number greater than 3" and "rolling an even number" would be mutually exclusive.

   Answer: F Type: T Sect: 8 Obj: 3 RANDOM: Y

7. A jar contains 5 white marbles, 6 red marbles, and 3 black marbles. If a marble is selected at random, the probability that it is white or black would be 4/7.

   Answer: T Type: T Sect: 8 Obj: 3 RANDOM: Y

8. Drawing a card from a deck and getting an ace, replacing it, and drawing another ace would be an example of independent events.

   Answer: T Type: T Sect: 7 Obj: 4 RANDOM: Y

9. A card is drawn from a deck and is not replaced. A second card is then drawn. The probability that both cards are nines would be 1/169.

   Answer: F Type: T Sect: 7 Obj: 4 RANDOM: Y

10. The probability that Janice is going over 70 miles per hour and get a ticket is 0.2. The probability that Janice is going over 70 mph is 0.4. The probability that Janice gets a ticket given that she is going over 70 mph would be 0.5.

    Answer: T Type: T   Sect: 7 Obj: 5   RANDOM: Y

11. A die is rolled. The complement of the event "rolling an odd number" would be 1, 3, and 5.

    Answer: F Type: T   Sect: 5 Obj: 6   RANDOM: Y

12. If a coin is tossed 4 times, the probability of getting at least 1 head would be 15/16.

    Answer: T Type: T   Sect: 5 Obj: 6   RANDOM: Y

MULTIPLE CHOICE

13. A drawer contains 12 black socks, 10 brown socks, and 6 white socks. If a sock is selected at random, what is the probability that it is white?
    a. 5/14   b. 3/7   c. 6/7   d. 3/14

    Answer: d Type: M   Sect: 2 Obj: 2   RANDOM: Y

14. A family has 4 children. How many possible outcomes will be in the sample space?
    a. 4
    b. 8
    c. 16
    d. 32

    Answer: c Type: M   Sect: 2 Obj: 2   RANDOM: Y

15. In a recent survey 8 people watched ABC, 7 people watched NBC, and 12 people watched CBS. If a person is picked at random, what is the probability that the person watched NBC or CBS?
    a. 19/27
    b. 20/27
    c. 5/9
    d. 7/27

    Answer: a Type: M   Sect: 2 Obj: 2   RANDOM: Y

16. If a die is rolled one time, what is the probability of getting a number less than 5 or an even number?
   a. 7/6
   b. 5/6
   c. 2/3
   d. 1

Answer: b Type: M   Sect: 8 Obj: 3   RANDOM: Y

17. Two dice are rolled. What is the probability of getting a 4 or 10?
   a. 11/36
   b. 1/18
   c. 1/3
   d. 1/6

Answer: d Type: M   Sect: 8 Obj: 3   RANDOM: Y

18. A card is drawn from an ordinary deck. What is the probability of getting a black jack?
   a. 1/26
   b. 1/52
   c. 1/13
   d. 3/52

Answer: a Type: M   Sect: 2 Obj: 2   RANDOM: Y

19. A single card is drawn from a deck. What is the probability of selecting a queen or a heart?
   a. 17/52
   b. 4/13
   c. 15/52
   d. 9/26

Answer: b Type: M   Sect: 8 Obj: 3   RANDOM: Y

20. The probability that a person will get a cold this year is 0.7. If 4 people are selected at random, what is the probability that all 4 will get a cold this year?
   a. 2.8
   b. 0.28
   c. 0.2401
   d. 0.02401

Answer: c Type: M   Sect: 7 Obj: 4   RANDOM: Y

21. If a single die is rolled, what is $P(4 \mid \text{even number})$?
   a. 1/2   b. 1/3   c. 1/6   d. 1

Answer: b Type: M   Sect: 6 Obj: 5   RANDOM: Y

22. The probability that Bob and Chris go to the movies is 0.35. The probability that Chris goes to the movies is 0.5. What is the probability that Bob goes to the movies given that Chris goes?
    a. 0.7
    b. 0.175
    c. 0.15
    d. 0.85

    Answer: a Type: M   Sect: 7 Obj: 5   RANDOM: Y

23. If a family has 5 children, what is the probability that at least one child is a girl?
    a. 1/2
    b. 31/32
    c. 15/16
    d. 29/32

    Answer: b Type: M   Sect: 5 Obj: 6   RANDOM: Y

24. For a single draw from a standard deck of cards, what is $P(\text{diamond}|\text{face card})$?
    a. 1/3   b. 3/13   c. 1/4   d. 1/2

    Answer: c Type: M   Sect: 6 Obj: 5   RANDOM: Y

25. For a single draw from a standard deck of cards, what is $P(\text{heart}|\text{red card})$?
    a. 1/4   b. 1/2   c. 1   d. 1/3

    Answer: b Type: M   Sect: 6 Obj: 5   RANDOM: Y

26. For a single draw from a standard deck of cards, what is $P(\text{queen}|\text{black card})$?
    a. 1/13   b. 2/13   c. 1/2   d. 1/4

    Answer: a Type: M   Sect: 6 Obj: 5   RANDOM: Y

27. Two dice are rolled. What is $P(8|\text{a roll of a double})$?
    a. 1/3   b. 1/5   c. 5/36   d. 1/6

    Answer: d Type: M   Sect: 6 Obj: 5   RANDOM: Y

28. One die is rolled. What is $P(5|\text{even number})$?
    a. 0   b. 1   c. 1/3   d. 1/6

    Answer: a Type: M   Sect: 6 Obj: 5   RANDOM: Y

SHORT ANSWER

29. A die is rolled. What is the probability of rolling a 5?

    Answer: 1/6   Type: S   Sect: 2 Obj: 2   RANDOM: Y

30. A box contains 30 red marbles, 10 blue marbles, and 5 yellow marbles. A marble is picked at random. What is the probability that a red marble was not selected?

    Answer: 1/3 Type: S   Sect: 5 Obj: 6   RANDOM: Y

31. Suppose you hold 30 tickets in a drawing for a color television. There were 500 tickets sold in the drawing. What is the probability that you win the TV?

    Answer: 3/50 Type: S   Sect: 2 Obj: 2   RANDOM: Y

32. A club has 8 doctors, 10 engineers, and 12 lawyers. If a person is picked at random, what is the probability that he or she is a lawyer?

    Answer: 2/5 Type: S   Sect: 2 Obj: 2   RANDOM: Y

33. A club has 8 doctors, 10 engineers, and 12 lawyers. If a person is picked at random, what is the probability that he or she is not an engineer?

    Answer: 2/3 Type: S   Sect: 5 Obj: 6   RANDOM: Y

34. If a die is rolled once, what is the probability of getting a 7?

    Answer: 0 Type: S   Sect: 2 Obj: 2   RANDOM: Y

35. If a die is rolled once, what is the probability of getting a number less than 3?

    Answer: 1/3 Type: S   Sect: 2 Obj: 2   RANDOM: Y

36. If a die is rolled once, what is the probability of getting a number greater than or equal to 5?

    Answer: 1/3 Type: S   Sect: 2 Obj: 2   RANDOM: Y

37. If a die is rolled once, what is the probability of getting a number more than 2 or an even number?

    Answer: 5/6 Type: S   Sect: 8 Obj: 3   RANDOM: Y

38.A certain type of car has a 10% chance of breaking down. Out of the next 90 cars, how many will break down?

Answer: 9 Type: S   Sect: 2 Obj: 2   RANDOM: Y

39.Two dice are rolled. What is the probability of getting a sum of 8?

Answer: 5/36 Type: S   Sect: 2 Obj: 2   RANDOM: Y

40.Two dice are rolled. What is the probability of getting a sum greater than 3?

Answer: 11/12 Type: S   Sect: 2 Obj: 2   RANDOM: Y

41.Two dice are rolled. What is the probability of getting a sum between 5 and 9?

Answer: 4/9 Type: S   Sect: 2 Obj: 2   RANDOM: Y

42.Two dice are rolled. What is the probability of getting a sum of 5 or doubles?

Answer: 5/18 Type: S   Sect: 8 Obj: 3   RANDOM: Y

43.A box contains 9 blue marbles, 11 red marbles, and 3 yellow marbles. If a marble is selected at random, what is the probability that it is yellow?

Answer: 3/23 Type: S   Sect: 2 Obj: 2   RANDOM: Y

44.A box contains 9 blue marbles, 11 red marbles, and 3 yellow marbles. If a marble is selected at random, what is the probability that it is not red?

Answer: 12/23 Type: S   Sect: 5 Obj: 6   RANDOM: Y

45.A box contains 9 blue marbles, 11 red marbles, and 3 yellow marbles. If a marble is selected at random, what is the probability that it is blue or red?

Answer: 20/23 Type: S   Sect: 8 Obj: 3   RANDOM: Y

46.One card is drawn from an ordinary deck of playing cards. What is the probability of getting a 5?

Answer: 1/13 Type: S   Sect: 2 Obj: 2   RANDOM: Y

47.One card is drawn from an ordinary deck of playing cards. What is the probability of getting an 8 or a jack?

Answer: 2/13 Type: S   Sect: 8 Obj: 3   RANDOM: Y

48. One card is drawn from an ordinary deck of playing cards. What is the probability of getting a black 9?

   Answer: 1/26 Type: S   Sect: 2 Obj: 2   RANDOM: Y

49. One card is drawn from an ordinary deck of playing cards. What is the probability of getting the king of hearts?

   Answer: 1/52 Type: S   Sect: 2 Obj: 2   RANDOM: Y

50. One card is drawn from an ordinary deck of playing cards. What is the probability of getting a diamond or a 6?

   Answer: 4/13 Type: S   Sect: 8 Obj: 3   RANDOM: Y

51. One card is drawn from an ordinary deck of playing cards. What is the probability of getting a card between 3 and 9?

   Answer: 5/13 Type: S   Sect: 2 Obj: 2   RANDOM: Y

52. In a dice game, a player loses if 4, 5, or 11 is tossed. What is the probability of losing on the first toss?

   Answer: 1/4 Type: S   Sect: 2 Obj: 2   RANDOM: Y

53. A couple plans to have 4 children. What is the probability of having all 4 girls?

   Answer: 1/16 Type: S   Sect: 2 Obj: 2   RANDOM: Y

54. If the probability that it will rain today is 0.3, what is the probability that it will not rain today?

   Answer: 0.7 Type: S   Sect: 5 Obj: 6   RANDOM: Y

55. A couple plans to have 4 children. What is the probability of having no girls?

   Answer: 1/16 Type: S   Sect: 2 Obj: 2   RANDOM: Y

56. Jim get A's in 85% of his courses. If he has taken 20 courses, how many A's has the received?

   Answer: 17 Type: S   Sect: 2 Obj: 2   RANDOM: Y

57. The SAT math scores for students at a small college are as follows:

Score	Percent
200-299	9%
300-399	29%
400-499	34%
500-599	19%
600-699	6%
700-799	3%

If a student is selected at random, what is the probability that the student's SAT score is over 599?

Answer: 9% or 0.09 Type: S   Sect: 2 Obj: 2   RANDOM: Y

58. The SAT math scores for students at a small college are as follows:

Score	Percent
200-299	9%
300-399	29%
400-499	34%
500-599	19%
600-699	6%
700-799	3%

If a student is selected at random, what is the probability that the student's SAT score is between 300 and 399?

Answer: 29% or 0.29 Type: S   Sect: 2 Obj: 2   RANDOM: Y

59. The SAT math scores for students at a small college are as follows:

Score	Percent
200-299	9%
300-399	29%
400-499	34%
500-599	19%
600-699	6%
700-799	3%

If a student is selected at random, what is the probability that the student's SAT score is at most 499?

Answer: 72% or 0.72 Type: S   Sect: 2 Obj: 2   RANDOM: Y

60. The SAT math scores for students at a small college are as follows:

Score	Percent
200-299	9%
300-399	29%
400-499	34%
500-599	19%
600-699	6%
700-799	3%

If a student is selected at random, what is the probability that the student's SAT score is between 300 and 699?

Answer: 88% or 0.88   Type: S   Sect: 2 Obj: 2   RANDOM: Y

61. The SAT math scores for students at a small college are as follows:

Score	Percent
200-299	9%
300-399	29%
400-499	34%
500-599	19%
600-699	6%
700-799	3%

If a student is selected at random, what is the probability that the student's SAT score is not more than 299?

Answer: 9% or 0.09 Type: S   Sect: 2 Obj: 2   RANDOM: Y

62. A jar contains 8 red, 2 yellow, and 3 brown marbles. What is the probability that a marble selected at random is red or brown?

Answer: 11/13 Type: S   Sect: 8 Obj: 3   RANDOM: Y

63. A month of the year is picked at random. What is the probability that it is a month with 31 days or begins with the letter J?

Answer: 2/3 Type: S   Sect: 8 Obj: 3   RANDOM: Y

64. A card is drawn from an ordinary deck of playing cards. What is the probability that it is a heart or a card between 6 and 9?

Answer: 19/52 Type: S   Sect: 8 Obj: 3   RANDOM: Y

65. The probability that a person owns a microwave is 0.65. The probability that a person owns a washer is 0.85. The probability that a person owns both a microwave and a washer is 0.58. What is the probability that a person owns a washer or a microwave?

    Answer: 0.92 Type: S   Sect: 8 Obj: 3   RANDOM: Y

66. The probability that Herbert shaves with a straight razor is 0.8. The probability that he uses after-shave is 0.6. The probability that he shaves with a straight razor or uses after-shave is 0.84. What is the probability that he shaves with a straight razor <u>and</u> uses after-shave?

    Answer: 0.56 Type: S   Sect: 8 Obj: 3   RANDOM: Y

67. The probability that Mary will visit Walt Disney World is 0.35, and the probability that she will visit Epcot Center is 0.7. The probability that she will visit both places is 0.42. What is the probability that Mary visits Walt Disney World or Epcot Center?

    Answer: 0.63 Type: S   Sect: 8 Obj: 3   RANDOM: Y

68. The staff at a small company includes 2 secretaries, 40 factory workers, 7 specialists, 3 engineers, and 2 executives. If a person is selected at random, what is the probability that he or she is a factory worker or a specialist?

    Answer: 47/54 Type: S   Sect: 8 Obj: 3   RANDOM: Y

69. The staff at a small company includes 2 secretaries, 40 factory workers, 7 specialists, 3 engineers, and 2 executives. If a person is selected at random, what is the probability that he or she is an engineer or a secretary?

    Answer: 5/54 Type: S   Sect: 8 Obj: 3   RANDOM: Y

70. In a college algebra class there are 18 freshmen and 17 sophomores. Eight of the freshmen are females and 13 of the sophomores are males. If a student is selected at random, what is the probability of selecting a sophomore or a female?

    Answer: 5/7 Type: S   Sect: 8 Obj: 3   RANDOM: Y

71. In a college algebra class there are 18 freshmen and 17 sophomores. Eight of the freshmen are females and 13 of the sophomores are males. If a student is selected at random, what is the probability of selecting a freshman or a male?

    Answer: 31/35 Type: S   Sect: 8 Obj: 3   RANDOM: Y

72. In a college algebra class there are 18 freshmen and 17 sophomores. Eight of the freshmen are females and 13 of the sophomores are males. If a student is selected at random, what is the probability of selecting a female or a male?

Answer: 1 Type: S   Sect: 8 Obj: 3   RANDOM: Y

73. A store has 3 colors of appliances, white, almond, and black. The types of appliances are listed below.

	White	Almond	Black
Refrigerator	12	10	3
Washer	10	8	5
Dryer	11	11	4

If an appliance is selected at random, what is the probability that it is black or white?

Answer: 45/74 Type: S   Sect: 8 Obj: 3   RANDOM: Y

74. A store has 3 colors of appliances, white, almond, and black. The types of appliances are listed below.

	White	Almond	Black
Refrigerator	12	10	3
Washer	10	8	5
Dryer	11	11	4

If an appliance is selected at random, what is the probability that it is almond or a refrigerator?

Answer: 22/37 Type: S   Sect: 8 Obj: 3   RANDOM: Y

75. A store has 3 colors of appliances, white, almond, and black. The types of appliances are listed below.

	White	Almond	Black
Refrigerator	12	10	3
Washer	10	8	5
Dryer	11	11	4

If an appliance is selected at random, what is the probability that it is not black?

Answer: 31/37 Type: S   Sect: 5 Obj: 6   RANDOM: Y

76. A store has 3 colors of appliances, white, almond, and black. The types of appliances are listed below.

	White	Almond	Black
Refrigerator	12	10	3
Washer	10	8	5
Dryer	11	11	4

If an appliance is selected at random, what is the probability that it is a washer or a black appliance?

Answer: 15/37 Type: S   Sect: 8 Obj: 3   RANDOM: Y

77. The frequency distribution shown here lists the number of cars looked at by customers in a dealer's lot on a Saturday.

No. of Cars	No. of Customers
1	4
2	18
3	25
4	10
5	8

If a customer is selected at random, what is the probability that the person looked at exactly 3 cars?

Answer: 5/13 Type: S   Sect: 2 Obj: 2   RANDOM: Y

78. The frequency distribution shown here lists the number of cars looked at by customers in a dealer's lot on a Saturday.

No. of Cars	No. of Customers
1	4
2	18
3	25
4	10
5	8

If a customer is selected at random, what is the probability that the person examined at most 2 cars?

Answer: 22/65 Type: S   Sect: 2 Obj: 2   RANDOM: Y

79. The frequency distribution shown here lists the number of cars looked at by customers in a dealer's lot on a Saturday.

No. of Cars	No. of Customers
1	4
2	18
3	25
4	10
5	8

If a customer is selected at random, what is the probability that the person examined at least 3 cars?

Answer: 43/65 Type: S   Sect: 2 Obj: 2   RANDOM: Y

80. There is a 45% chance that it will snow in Colorado Springs, a 60% chance it will snow in Denver, and a 35% chance that it will snow in both cities. What is the probability that it will snow in Denver or Colorado Springs?

Answer: 70% of 0.70 Type: S   Sect: 8 Obj: 3   RANDOM: Y

81. The probability that General Motors stock will fall is 0.2, the probability that AT&T stock will fall is 0.1, and the probability that both will fall is 0.05. What is the probability that GM or AT&T stock will fall?

Answer: 0.25 Type: S   Sect: 8 Obj: 3   RANDOM: Y

82. The probability that Joyce owns a VCR is 0.8, the probability that she owns a CD player is 0.6, and the probability that she owns a VCR or a CD player is 0.98. What is the probability that she owns both a VCR and a CD player?

Answer: 0.42 Type: S   Sect: 8 Obj: 3   RANDOM: Y

83. In a recent study, the following data were obtained in response to the question "Do you favor recycling in your neighborhood?"

	Yes	No	No Opinion
Males	40	15	5
Females	30	20	10

If a person is picked at random, what is the probability that the person is in favor of recycling?

Answer: 7/12 Type: S Sect: 2 Obj: 2   RANDOM: Y

84. In a recent study, the following data were obtained in response to the question "Do you favor recycling in your neighborhood?"

	Yes	No	No Opinion
Males	40	15	5
Females	30	20	10

If a person is picked at random, what is the probability that the person is male or does not favor recycling?

Answer: 2/3 Type: S   Sect: 8 Obj: 3   RANDOM: Y

85. In a recent study, the following data were obtained in response to the question "Do you favor recycling in your neighborhood?"

	Yes	No	No Opinion
Males	40	15	5
Females	30	20	10

If a person is picked at random, what is the probability that the person is male and is in favor of recycling?

Answer: 1/3 Type: S   Sect: 8 Obj: 3   RANDOM: Y

86. If one card is drawn from an ordinary deck of cards, what is the probability that it is a club or a heart or a king?

Answer: 7/13 Type: S   Sect: 8 Obj: 3   RANDOM: Y

87. If one card is drawn from an ordinary deck of cards, what is the probability that it is an 8 or a queen or a spade?

Answer: 19/52 Type: S   Sect: 8 Obj: 3   RANDOM: Y

88. If one card is drawn from an ordinary deck of cards, what is the probability that it is a club or a diamond or a heart?

Answer: 3/4 Type: S   Sect: 8 Obj: 3   RANDOM: Y

89. If one card is drawn from an ordinary deck of cards, what is the probability that it is a red card or a 10?

Answer: 7/13 Type: S   Sect: 8 Obj: 3   RANDOM: Y

90. If one card is drawn from an ordinary deck of cards, what is the probability that it is a diamond or a king?

Answer: 4/13 Type: S   Sect: 8 Obj: 3   RANDOM: Y

91. If one card is drawn from an ordinary deck of cards, what is the probability that it is a club or a red jack?

Answer: 15/52 Type: S   Sect: 8 Obj: 3   RANDOM: Y

92. There is a 76% chance that a man will give a Valentine Day card to his wife. If 4 men are selected at random, what is the probability that all 4 will give valentines to their wives?

Answer: 0.334 Type: S   Sect: 7 Obj: 4   RANDOM: Y

93. If 32% of the families surveyed had a cat, what is the probability that all 3 of the next families surveyed have a cat?

Answer: 0.033 Type: S   Sect: 7 Obj: 4   RANDOM: Y

94. A student takes a 4 question multiple-choice quiz by guessing. Each question has 4 choices. What is the probability that the student misses all 4 questions?

Answer: 0.316 or 81/256 Type: S   Sect: 7 Obj: 4   RANDOM: Y

95. What is the probability that an archer with an 80% accuracy rate will hit a bull's-eye 5 times in a row?

Answer: 0.328 Type: S   Sect: 7 Obj: 4   RANDOM: Y

96. A card is drawn and replaced, then a second card is drawn. What is the probability of getting 2 clubs?

Answer: 1/16 Type: S   Sect: 7 Obj: 4   RANDOM: Y

97. An urn contains 6 green balls, 5 white balls, and 3 brown balls. A ball is selected, its color is noted, and then it is replaced. What is the probability of selecting 2 brown balls?

Answer: 9/196 Type: S   Sect: 7 Obj: 4   RANDOM: Y

98. An urn contains 6 green balls, 5 white balls, and 3 brown balls. A ball is selected, its color is noted, and then it is replaced. What is the probability of selecting a green ball, then a white ball?

    Answer: 15/98 Type: S   Sect: 7 Obj: 4   RANDOM: Y

99. An urn contains 6 green balls, 5 white balls, and 3 brown balls. A ball is selected, its color is noted, and then it is replaced. What is the probability of selecting a brown ball, then a green ball?

    Answer: 9/98 Type: S   Sect: 7 Obj: 4   RANDOM: Y

100. An urn contains 6 blue balls, 8 red balls, 7 green balls and 2 white balls. What is the probability that a green ball is not selected, if we select one ball at random?

    Answer: 16/23 Type: S   Sect: 5 Obj: 6   RANDOM: Y

101. The probability that a medical test will show positive is 0.27. If 3 people are tested, what is the probability that all 3 show positive?

    Answer: 0.0197 Type: S   Sect: 7 Obj: 4   RANDOM: Y

102. If 4 people are selected at random, what is the probability that all were born on the same day of the week?

    Answer: 1/2401 Type: S   Sect: 7 Obj: 4   RANDOM: Y

103. A coin is tossed 3 times in a row. What is the probability that all 3 tosses are tails?

    Answer: 1/8 Type: S   Sect: 7 Obj: 4   RANDOM: Y

104. A box contains 9 red, 5 yellow, and 4 blue marbles. A marble is drawn, it is not replaced, and a second marble is drawn. What is the probability of selecting 2 yellow marbles?

    Answer: 10/153 Type: S   Sect: 7 Obj: 4   RANDOM: Y

105. A box contains 9 red, 5 yellow, and 4 blue marbles. A marble is drawn, it is not replaced, and a second marble is drawn. What is the probability of selecting a blue, then a red?

    Answer: 2/17 Type: S   Sect: 7 Obj: 4   RANDOM: Y

106. A card is drawn from a standard deck, it is replaced, and then a second card is drawn. What is the probability of drawing two diamonds?

    Answer: 1/16 Type: S   Sect: 7 Obj: 4   RANDOM: Y

107. A box contains 9 red, 5 yellow, and 4 blue marbles. A marble is drawn, it is not replaced, and a second marble is drawn. What is the probability of selecting no blue marbles?

    Answer: 91/153 Type: S   Sect: 7 Obj: 4   RANDOM: Y

108. Three cards are drawn from a deck without replacement. What is the probability of selecting 3 eights?

    Answer: 1/5525 Type: S   Sect: 7 Obj: 4   RANDOM: Y

109. Three cards are drawn from a deck with replacement. What is the probability of selecting a heart and 2 diamonds?

    Answer: 1/64 Type: S   Sect: 7 Obj: 4   RANDOM: Y

110. Three cards are drawn from a deck with replacement. What is the probability of selecting a king, a 9, and a 10?

    Answer: 1/2197 Type: S   Sect: 7 Obj: 4   RANDOM: Y

111. Three cards are drawn from a deck with replacement. What is the probability of selecting 2 black cards and a heart?

    Answer: 1/16 Type: S   Sect: 7 Obj: 4   RANDOM: Y

112. In a shipment of 30 televisions, 3 are defective. If 2 sets are randomly selected and tested, what is the probability that both are defective? (The first one is not replaced before the second one is selected.)

    Answer: 1/145 Type: S   Sect: 7 Obj: 4   RANDOM: Y

113. In a study of snails, 9 out of 10 are pregnant. If 4 snails are selected without replacement, what is the probability that all four are pregnant?

    Answer: 3/5 Type: S   Sect: 7 Obj: 4   RANDOM: Y

114. A pair of dice is rolled 3 times. What is the probability of getting a 7 on all three rolls?

    Answer: 1/216 Type: S   Sect: 7 Obj: 4   RANDOM: Y

115. The probability that a student gets an A in both calculus and chemistry is 0.15. The probability that a student get an A in calculus is 0.2. What is the probability that a student gets an A in chemistry given that he or she gets an A in calculus?

     Answer: 0.75 Type: S   Sect: 7 Obj: 5   RANDOM: Y

116. In a subdivision, 80% of the homes have garages and 45% have both a garage and a patio. If a home is selected at random, what is the probability that the home has a patio given that it has a garage?

     Answer: 9/16 or 0.5625 Type: S   Sect: 7 Obj: 5   RANDOM: Y

117. The probability that a plane will leave on time is 0.8. The probability that it will leave on time and also arrive on time is 0.7. What is the probability that a plane will arrive on time given that it leaves on time?

     Answer: 7/8 or 0.875 Type: S   Sect: 7 Obj: 5   RANDOM: Y

118. A study of the faculty at a small college was conducted to determine their number of years of experience. The results are as follows:

Experience	Instructors	Assistant Professors	Professors
1-5 yrs.	8	3	2
6-10 yrs.	4	15	10
11-15 yrs.	1	12	8

     If a faculty member is selected at random, what is the probability the person has 6-10 years of experience given that he or she is a professor?

     Answer: 1/2 Type: S   Sect: 7 Obj: 5   RANDOM: Y

119. A study of the faculty at a small college was conducted to determine their number of years of experience. The results are as follows:

Experience	Instructors	Assistant Professors	Professors
1-5 yrs.	8	3	2
6-10 yrs.	4	15	10
11-15 yrs.	1	12	8

     If a faculty member is selected at random, what is the probability the person is an assistant professor given that he or she has 11-15 years of experience?

     Answer: 4/7 Type: S   Sect: 7 Obj: 5   RANDOM: Y

120. A study of the faculty at a small college was conducted to determine their number of years of experience. The results are as follows:

Experience	Instructors	Assistant Professors	Professors
1-5 yrs.	8	3	2
6-10 yrs.	4	15	10
11-15 yrs.	1	12	8

If a faculty member is selected at random, what is the probability the person is a full professor and has 6-10 years of experience?

Answer: 10/63 Type: S   Sect: 7 Obj: 4   RANDOM: Y

121. A study of the faculty at a small college was conducted to determine their number of years of experience. The results are as follows:

Experience	Instructors	Assistant Professors	Professors
1-5 yrs.	8	3	2
6-10 yrs.	4	15	10
11-15 yrs.	1	12	8

If a faculty member is selected at random, what is the probability the person is an instructor given that he or she has 6-10 years of experience?

Answer: 4/29 Type: S   Sect: 7 Obj: 5   RANDOM: Y

122. Sixty students enter a Ping-Pong tournament. They are classified by gender and class in school. The results are in the table.

	Juniors	Seniors
Males	25	20
Females	10	5

If a student is selected at random, what is the probability the student is a male given that the person is a senior?

Answer: 4/5 Type: S   Sect: 7 Obj: 5   RANDOM: Y

123. Sixty students enter a Ping-Pong tournament. They are classified by gender and class in school. The results are in the table.

	Juniors	Seniors
Males	25	20
Females	10	5

If a student is selected at random, what is the probability the student is a junior given that the person is a female?

Answer: 2/3 Type: S   Sect: 7 Obj: 5   RANDOM: Y

124. Sixty students enter a Ping-Pong tournament. They are classified by gender and class in school. The results are in the table.

	Juniors	Seniors
Males	25	20
Females	10	5

If a student is selected at random, what is the probability the student is a male or a junior?

Answer: 11/12 Type: S   Sect: 8 Obj: 3   RANDOM: Y

125. Sixty students enter a Ping-Pong tournament. They are classified by gender and class in school. The results are in the table.

	Juniors	Seniors
Males	25	20
Females	10	5

If a student is selected at random, what is the probability the student is a female given that the person is a junior?

Answer: 2/7 Type: S   Sect: 7 Obj: 5   RANDOM: Y

126. The following table lists ages of cars and trucks.

Ages	Cars	Trucks
1-3	20	15
4-6	35	30
7-9	10	25

If a vehicle is selected at random, what is the probability it is a truck given that it is 4-6 years old?

Answer: 6/13 Type: S   Sect: 7 Obj: 5   RANDOM: Y

127. The following table lists ages of cars and trucks.

Ages	Cars	Trucks
1-3	20	15
4-6	35	30
7-9	10	25

If a vehicle is selected at random, what is the probability it is a car given that it is 7-9 years old?

Answer: 2/7 Type: S   Sect: 7 Obj: 5   RANDOM: Y

128. The following table lists ages of cars and trucks.

Ages	Cars	Trucks
1-3	20	15
4-6	35	30
7-9	10	25

If a vehicle is selected at random, what is the probability it is a truck or is 4-6 years old?

Answer: 7/9 Type: S   Sect: 8 Obj: 3   RANDOM: Y

129. According to the weather forecaster, the probability that it will rain tomorrow is 0.7. What is the probability that it will not rain tomorrow?

Answer: 0.3 Type: S   Sect: 5 Obj: 6   RANDOM: Y

130. It has been suggested that 4% of all cars on the freeway are going over 70 miles per hour. If the speeds of 4 automobiles are checked, what is the probability that at least one car is going over 70 miles per hour?

Answer: 0.151 Type: S   Sect: 5 Obj: 6   RANDOM: Y

131. Let A = a diamond, B = a king, and C = a face card. A card is selected at random. What is $P(A$ and $B)$?

    Answer: 1/52 Type: S   Sect: 7 Obj: 4   RANDOM: Y

132. Let A = a diamond, B = a king, and C = a face card. A card is selected at random. What is $P(B$ and $C)$?

    Answer: 1/13 Type: S   Sect: 7 Obj: 4   RANDOM: Y

133. Let A = a diamond, B = a king, and C = a face card. A card is selected at random. What is $P(A$ or $C)$?

    Answer: 11/26 Type: S   Sect: 8 Obj: 3   RANDOM: Y

134. Let A = a diamond, B = a king, and C = a face card. A card is selected at random. What is $P(B$ or $C)$?

    Answer: 3/13 Type: S   Sect: 8 Obj: 3   RANDOM: Y

135. Let A = a diamond, B = a seven, and C = a face card. A card is selected at random. What is $P(A$ or $B)$?

    Answer: 4/13 Type: S   Sect: 8 Obj: 3   RANDOM: Y

136. The following table shows different types of schools in Jackson County with the number of students enrolled.

Level	Public	Private
Elementary	15,000	3,500
High School	10,000	1,500
College	18,000	1,200

If a student is selected at random, what is the probability that he or she attends a private school?

Answer: 6,200/49,200 = 31/246 Type: S   Sect: 2 Obj: 2   RANDOM: Y

137. The following table shows different types of schools in Jackson County with the number of students enrolled.

Level	Public	Private
Elementary	15,000	3,500
High School	10,000	1,500
College	18,000	1,200

If a student is selected at random, what is the probability that he attends an elementary school or a public school?

Answer: 465/492 = 155/164 Type: S   Sect: 2 Obj: 3   RANDOM: Y

138. If the probability that a robin eats at a bird feeder is 0.2, the probability that a blue jay eats at a bird feeder is 0.85, and the probability is .15 that both birds eat at the feeder, what is the probability that a robin or a blue jay eats at the feeder?

Answer: 0.9 Type: S   Sect: 8 Obj: 3   RANDOM: Y

139. The following table shows different types of schools in Jackson County with the number of students enrolled.

Level	Public	Private
Elementary	15,000	3,500
High School	10,000	1,500
College	18,000	1,200

If a student is selected at random, what is the probability that the student attends college given that he or she goes to a private school?

Answer: 12/62 = 6/31 Type: S   Sect: 7 Obj: 5   RANDOM: Y

140. The following table shows different types of schools in Jackson County with the number of students enrolled.

Level	Public	Private
Elementary	15,000	3,500
High School	10,000	1,500
College	18,000	1,200

If a student is selected at random, what is the probability that the student attends high school given that the student attends a public institution?

Answer: 10/43 Type: S   Sect: 7 Obj: 5   RANDOM: Y

141. The probability that a student gets an A is .10, the probability that he gets a B is .23, the probability that he get a C is .35, and the probability that he gets a D is 0.05. What is the probability that he will get an F?

     Answer: 0.27 Type: S   Sect: 5 Obj: 5   RANDOM: Y

142. If the probability is 0.03 that the post office will lose a letter that is sent, what is the probability that it will not lose a letter?

     Answer: 0.97 Type: S   Sect: 5 Obj: 5   RANDOM: Y

143. If the occurrence of one outcome precludes the occurrence of other outcomes on the same trial, what are these outcomes called?

     Answer: mutually exclusive Type: S   Sect: 1 Obj: 1   RANDOM: Y

144. What is the set of all possible outcomes of a statistical experiment called?

     Answer: a sample space Type: S   Sect: 1 Obj: 1   RANDOM: Y

145. What is the largest possible value of $P(A)$?

     Answer: 1 Type: S   Sect: 4 Obj: 1   RANDOM: Y

CHAPTER 5

TRUE/FALSE

1. One requirement for a probability distribution is that the sum of the probabilities of all the events in the sample space must equal one.

   Answer: T Type: T   Sect: 2 Obj: 1   RANDOM: Y

2. Listed below is a distribution.

x	1	2	3	4
$P(x)$	1/2	1/3	1/4	1/5

   This distribution is a probability distribution.

   Answer: F Type: T   Sect: 2 Obj: 1   RANDOM: Y

3. The number of hot dogs sold at a fair is an example of a discrete distribution.

   Answer: T Type: T   Sect: 2 Obj: 1   RANDOM: Y

4. Using the probability distribution listed, the mean would be 1.9.

x	0	1	2	3
$P(x)$	0.2	0.1	0.3	0.4

   Answer: T Type: T   Sect: 2 Obj: 3   RANDOM: Y

5. If the standard deviation of a probability distribution is 2.53, the variance is 1.59.

   Answer: F Type: T   Sect: 2 Obj: 3   RANDOM: Y

6. In binomial experiments, the outcomes are usually classified as successes or failures.

   Answer: T Type: T   Sect: 3 Obj: 4   RANDOM: Y

7. If a coin is tossed four times, the probability of getting exactly 3 tails would be 3/4.

   Answer: F Type: T   Sect: 3 Obj: 4   RANDOM: Y

8. The probability of getting a defective light bulb was 0.03. In a shipment of 1,500 light bulbs we would expect to get 45 defective ones.

   Answer: T Type: T   Sect: 5 Obj: 5   RANDOM: Y

9. A survey was conducted to determine what type of food a person preferred. There were four different responses. In order to find the probabilities, we would use the binomial distribution.

   Answer: F Type: T   Sect: 3 Obj: 6   RANDOM: Y

10. If a coin is tossed, the possible outcomes are true or false.

    Answer: F   Type: T   Sect: 1 Obj: 6   RANDOM: Y

11. If we can count the individual values the random variable can assume, we say the variable is discrete.

    Answer: T   Type: T   Sect: 1 Obj: 6   RANDOM: Y

MULTIPLE CHOICE

12. A person flips a coin 6 times. Let x be the number of tails. Which of the following values could not be a value of x for this experiment?
    a. 5
    b. 6
    c. 0
    d. 7

    Answer: d Type: M   Sect: 2 Obj: 1   RANDOM: Y

13. A probability distribution is constructed for the number of boys in a family that has 2 children. Let x be the number of boys. What is the probability for $x = 2$?
    a. 1/4
    b. 1/2
    c. 1/3
    d. 3/4

    Answer: a Type: M   Sect: 2 Obj: 1   RANDOM: Y

14. What value would be needed to complete the following probability distribution?

x	0	1	2	3	4
$P(x)$	1/3	1/8		1/4	1/6

    a. 1/5
    b. 1/12
    c. 1/24
    d. 1/8

    Answer: d Type: M   Sect: 2 Obj: 2   RANDOM: Y

15. What is the mean of the following probability distribution?

x	0	1	2	3	4
$P(x)$	0.2	0.1	0.35	0.05	0.3

   a. 2.05
   b. 2.15
   c. 1.95
   d. 2

Answer: b Type: M   Sect: 2 Obj: 3   RANDOM: Y

16. What is the standard deviation of the following probability distribution?

x	0	2	4	6	8
$P(x)$	0.25	0.1	0.3	0.25	0.1

   a. 4.37
   b. 2.63
   c. 3.7
   d. 6.91

Answer: b Type: M   Sect: 2 Obj: 3   RANDOM: Y

17. A student randomly guesses at 7 multiple-choice questions. Each question has 5 possible choices. What is the probability that the student gets exactly 3 correct?
   a. 0.1147
   b. 0.2753
   c. 0.0287
   d. 0.1730

Answer: a Type: M   Sect: 4 Obj: 4   RANDOM: Y

18. A binomial function has $n = 5$, $p = 0.35$, and $x = 2$. What is the probability associated with this data?
   a. 0.3124
   b. 0.3364
   c. 0.1811
   d. 0.3456

Answer: b Type: M   Sect: 4 Obj: 4   RANDOM: Y

19. If we have a binomial distribution with $n = 200$ and $p = 0.47$, what would be the variance?
   a. 49.82
   b. 94
   c. 7.06
   d. 9.70

Answer: a Type: M   Sect: 5 Obj: 5   RANDOM: Y

20. A binomial distribution has a mean of 4.5 with $n = 75$. What is p?
    a. 0.06   b. 0.16   c. 0.05   d. 337.5

    Answer: a Type: M   Sect: 5 Obj: 5   RANDOM: Y

21. A binomial distribution has $n = 400$ and $p = 0.08$. What is the standard deviation?
    a. 32   b. 29.44   c. 5.6569   d. 5.4259

    Answer: d Type: M   Sect: 5 Obj: 5   RANDOM: Y

22. Four out of nine people applying for a job at a bank have a bachelor's degree. If 360 people apply for a job at the bank, how many would you expect to have a bachelor's degree?
    a. 81   b. 160   c. 150   d. 200

    Answer: b   Type: M   Sect: 5 Obj: 5   RANDOM: Y

SHORT ANSWER

Determine if the following distribution represents a probability distribution. If not, state why not.

23.

x	1	3	5	7
$P(x)$	1/3	1/3	1/3	1/3

Answer: Not a probability distribution. The sum of the probabilities is more than one.

Type: S   Sect: 2 Obj: 1   RANDOM: Y

24.

x	0	1	2	3
$P(x)$	0.2	0.5	0.3	-0.1

Answer: Not a probability distribution. A probability cannot be negative. Type: S

Sect: 2 Obj: 1   RANDOM: Y

25.

x	5	7	9	11	12
$P(x)$	0.2	0.3	0.2	0.1	0.2

Answer: It is a probability distribution. Type: S   Sect: 2 Obj: 1   RANDOM: Y

26. $P(x) = x^2/36$ for $x = 1, 2, 3, 4$

    Answer: Not a probability distribution. It does not add up to one. Type: S   Sect: 2

    Obj: 1   RANDOM: Y

27. $P(x) = \dfrac{2x - 1}{15}$ for $x = 1, 3, 5$

    Answer: It is a probability distribution.   Type: S   Sect: 2 Obj: 1   RANDOM: Y

28. $P(x) = 0.25$ for $x = 0, 1, 2, 3$

    Answer: It is a probability distribution.   Type: S   Sect: 2 Obj: 1   RANDOM: Y

29. $P(x) = \dfrac{|x - 3|}{9}$ for $x = 0, 1, 2, 3, 4,$ and 5

    Answer: It is a probability distribution.   Type: S   Sect: 2 Obj: 1   RANDOM: Y

Construct a probability distribution for the given data.

30. The probabilities that a college student will get 0, 1, 2, 3, or 4 A's at the end of the semester are 0.15, 0.3, 0.25, 0.2, and 0.1, respectively.

    Answer:

x	0	1	2	3	4
$P(x)$	0.15	0.3	0.25	0.2	0.1

    Type: S   Sect: 2 Obj: 2   RANDOM: Y

31. The probabilities that a customer has to wait 2, 3, 4, or 5 minutes in the checkout line are 1/3, 1/4, 1/6, and 1/4, respectively.

    Answer:

x	2	3	4	5
$P(x)$	1/3	1/4	1/6	1/4

    Type: S   Sect: 2 Obj: 2   RANDOM: Y

32. A box contains 5 slips of paper with the number 1 on it, 4 slips of paper with the number 5 on it, and 10 slips of paper with the number 10 on it. Construct a probability distribution for this data.

    Answer:

x	1	5	10
$P(x)$	5/19	4/19	10/19

    Type: S   Sect: 2 Obj: 2   RANDOM: Y

33. Construct a probability distribution for tossing 4 coins. Let x represent the number of tails appearing in each toss.

    Answer:

x	0	1	2	3	4
$P(x)$	1/16	1/4	3/8	1/4	1/16

    Type: S   Sect: 2 Obj: 2   RANDOM: Y

34. If 400 tickets are sold at $2.00 each for a $250 VCR, and a person purchases 1 ticket, what is the expected value?

    Answer: -$1.375 Type: S   Sect: 2 Obj: 3   RANDOM: Y

35. A farmer loses $50,000 when the summer is too dry and makes $100,000 when the growing season is moderate. The probability of the growing season being too dry is 35%. What is the expectation for profit?

    Answer: $47,500 Type: S   Sect: 2 Obj: 3   RANDOM: Y

36. For a raffle, 2,500 tickets are sold for $1.00 each for 4 cash prizes of $250, $100, $50, and $25. What is the expected value if a person purchases 5 tickets?

    Answer: -$4.15 Type: S   Sect: 2 Obj: 3   RANDOM: Y

37. If a player rolls a 5 or a 7 he wins $10. The cost to play the game is $2.00. What is the expectation of the game?

    Answer: $1.33 Type: S   Sect: 2 Obj: 3   RANDOM: Y

38. If a person selects a face card from a deck of cards, he will win $5.00. If the game is to be fair, how much should he pay to play the game?

    Answer: $1.50 Type: S   Sect: 2 Obj: 3   RANDOM: Y

Find the mean for the following probability distribution.

39.

No. of Customers Waiting	1	2	3	4	5
Probability $P(x)$	0.15	0.32	0.28	0.15	0.10

    Answer: 2.73 Type: S   Sect: 2 Obj: 3   RANDOM: Y

40.

No. of Glasses of Milk	0	1	2	3	4
Probability $P(x)$	0.18	0.33	0.27	0.16	0.06

Answer: 1.59 Type: S   Sect: 2 Obj: 3   RANDOM: Y

41.

No. of Recliners Sold	2	4	6	8	10
Probability $P(x)$	0.20	0.35	0.23	0.17	0.05

Answer: 5.04 Type: S   Sect: 2 Obj: 3   RANDOM: Y

Find the variance and standard deviation for the given distribution.

42.

No. of Credit Cards Owned	0	1	2	3	4
Probability $P(x)$	0.2	0.15	0.3	0.3	0.05

Answer: variance = 1.4275, standard deviation = 1.19 Type: S   Sect: 2 Obj: 3

RANDOM: Y

43.

No. of TV's per Household	1	2	3	4
Probability $P(x)$	0.4	0.3	0.2	0.1

Answer: variance = 1, standard deviation = 1 Type: S   Sect: 2 Obj: 3   RANDOM: Y

44. The numbers 0, 1, 2, 3, and 4 were put in a paper bag and drawn out a number of times. The probability distribution for the occurrence of each number is listed below.

x	0	1	2	3	4
$P(x)$	1/4	1/8	1/8	1/3	1/6

Answer: variance = 2.12, standard deviation = 1.46 Type: S   Sect: 2 Obj: 3

RANDOM: Y

45.

No. of Customers Waiting	1	2	3	4	5
Probability $P(x)$	0.15	0.32	0.28	0.15	0.10

Answer: variance = 1.397, standard deviation = 1.182 Type: S   Sect: 2 Obj: 3

RANDOM: Y

46.

No. of Glasses of Milk	0	1	2	3	4
Probability $P(x)$	0.18	0.33	0.27	0.16	0.06

Answer: variance = 1.282, standard deviation = 1.132 Type: S   Sect: 2 Obj: 3

RANDOM: Y

47.

No. of Recliners Sold	2	4	6	8	10
Probability $P(x)$	0.20	0.35	0.23	0.17	0.05

Answer: variance = 5.158, standard deviation = 2.271   Type: S   Sect: 2   Obj: 3

RANDOM: Y

Use the binomial formula to compute the following probability.

48. If $n = 6$, $p = 2/3$, and $x = 4$, what is the probability?

Answer: 0.329 Type: S   Sect: 3 Obj: 4   RANDOM: Y

49. If $n = 4$, $p = 1/8$, and $x = 2$, what is the probability?

Answer: 0.0718 Type: S   Sect: 3 Obj: 4   RANDOM: Y

50. If $n = 7$, $p = 0.65$, and $x = 5$, what is the probability?

Answer: 0.2985 Type: S   Sect: 4 Obj: 4   RANDOM: Y

51. If $n = 8$, $p = 0.07$, and $x = 1$, what is the probability?

Answer: 0.337 Type: S   Sect: 4 Obj: 4   RANDOM: Y

52. A student takes a 9-question multiple-choice test by guessing. Each question has 5 choices. What is the probability that at least 4 guesses are correct?

Answer: 0.0857 Type: S   Sect: 4 Obj: 4   RANDOM: Y

53. A student takes a 9-question multiple-choice test by guessing. Each question has 5 choices. What is the probability that at most 3 guesses are correct?

Answer: 0.9144 Type: S   Sect: 4 Obj: 4   RANDOM: Y

54. A student takes a 9-question multiple-choice test by guessing. Each question has 5 choices. What is the probability that exactly 2 guesses will be correct?

Answer: 0.302 Type: S   Sect: 4 Obj: 4   RANDOM: Y

55. A student takes a 9-question multiple-choice test by guessing. Each question has 5 choices. What is the probability that less than 2 guesses will be correct?

Answer: 0.4362 Type: S   Sect: 4 Obj: 4   RANDOM: Y

56. If 60% of all households own VCRs, what is the probability that if 11 households are selected at random, more than 7 will own VCRs?

Answer: 0.2963 Type: S   Sect: 4 Obj: 4   RANDOM: Y

57. If 60% of all households own VCRs, what is the probability that if 11 households are selected at random, at most 8 will own VCRs?

Answer: 0.8811 Type: S   Sect: 4 Obj: 4   RANDOM: Y

58. If 60% of all households own VCRs, what is the probability that if 11 households are selected at random, exactly 6 will own VCRs?

Answer: 0.2207 Type: S   Sect: 4 Obj: 4   RANDOM: Y

59. Mary says that only 10% of her 15 classmates remember who ran for president in 1968. What is the probability that exactly 2 remember who ran for president in 1968?

Answer: 0.2669 Type: S   Sect: 4 Obj: 4   RANDOM: Y

60. In a large metropolitan area 30% of the homes have 2 bedrooms. What is the probability that in a sample of 14 homes, more than 4 have 2 bedrooms?

Answer: 0.4157 or 0.4158 Type: S   Sect: 4 Obj: 4   RANDOM: Y

61. In a large metropolitan area 30% of the homes have 2 bedrooms. What is the probability that in a sample of 14 homes, at most two have 2 bedrooms?

Answer: 0.1609 Type: S   Sect: 4 Obj: 4   RANDOM: Y

62. In a large metropolitan area 30% of the homes have 2 bedrooms. What is the probability that in a sample of 14 homes, exactly five have 2 bedrooms?

Answer: 0.1963 Type: S   Sect: 4 Obj: 4   RANDOM: Y

63. In a large metropolitan area 30% of the homes have 2 bedrooms. If there were 150,000 homes in the area, how many would you expect to find with 2 bedrooms?

Answer: 45,000 Type: S   Sect: 5 Obj: 5   RANDOM: Y

64. A beauty operator estimates that 20% of her customers want a permanent on any given day. What is the probability that out of her next 15 customers more than 4 want a permanent?

Answer: 0.1643 Type: S   Sect: 4 Obj: 4   RANDOM: Y

65. An encyclopedia salesman makes sales to approximately 15% of the people he contacts. If he contacts 500 people this month, how many would you expect to purchase encyclopedias from this salesman?

    Answer: 75 Type: S   Sect: 5 Obj: 5   RANDOM: Y

66. The probability that Franklin makes a free throw is 60%. If he shoots 18 times, what is the probability that he makes at most 8 free throws?

    Answer: 0.1348 Type: S   Sect: 4 Obj: 4   RANDOM: Y

67. The foreman at a large plant estimates that parts are defective about 1% of the time. If the plant produces 25,000 parts in a week, what are the mean, variance, and standard deviation for the number of defective parts?

    Answer: mean = 250, variance = 247.5, standard deviation = 15.73 Type: S   Sect: 5

    Obj: 5   RANDOM: Y

68. The failure rate for taking the bar exam in Central City is 35%. If 250 people take the bar exam, what are the mean, variance, and standard deviation for the number of failures?

    Answer: mean = 87.5, variance = 56.875, standard deviation = 7.54 Type: S   Sect: 5

    Obj: 5   RANDOM: Y

69. A certain species of fish lays 300 eggs. The survival rate is predicted to be 58%. What are the mean and standard deviation for the number of fish that will survive?

    Answer: the mean is 174 and the standard deviation is 8.55 Type: S   Sect: 5 Obj: 5

    RANDOM: Y

70. Jim draws a card from a deck, replaces it, and draws again. He draws 50 times. What are the mean and standard deviation for the number of kings he would expect to draw?

    Answer: mean = 3.85, standard deviation = 1.88 Type: S   Sect: 5 Obj: 5

    RANDOM: Y

71. At a bridge tournament, it was found that 43% of the players smoked. If 4,000 people were at the tournament, what are the mean and standard deviation for those players who do not smoke?

    Answer: mean = 2,280, standard deviation = 31.31 Type: S Sect: 5 Obj: 5

    RANDOM: Y

72. At a certain university, it was found that only 29% of those who enter as freshmen
    actually graduate. If 5,500 enter as freshmen, how many actually graduate?

    Answer: 1,595 Type: S   Sect: 5 Obj: 5   RANDOM: Y

73. A tire manufacturer knows that about 3% of its tires are defective. If 12 tires are
    selected at random, what is the probability that at most 2 are defective?

    Answer: 0.9951 Type: S   Sect: 4 Obj: 4   RANDOM: Y

74. A tire manufacturer knows that about 3% of its tires are defective. If 12 tires are
    selected at random, what is the probability that at least one tire is defective?

    Answer: 0.3062 Type: S   Sect: 4 Obj: 4   RANDOM: Y

75. A tire manufacturer knows that about 3% of its tires are defective. If 12 tires are
    selected at random, what is the probability that exactly 2 tires are defective?

    Answer: 0.0438 Type: S   Sect: 4 Obj: 4   RANDOM: Y

76. A tire manufacturer knows that about 3% of its tires are defective. If 12 tires are
    selected at random, what is the probability that exactly 9 are good tires?

    Answer: 0.0045 Type: S   Sect: 4 Obj: 4   RANDOM: Y

77. The probability that a person will have 1 dental checkup per year is 0.5. If 9 people
    are selected at random, what is the probability that more than 6 will have 1 dental
    checkup this year?

    Answer: 0.0899 Type: S   Sect: 4 Obj: 4   RANDOM: Y

78. The probability that a person will have 1 dental checkup per year is 0.5. If nine
    people are selected at random, what is the probability that at least 4 will have 1
    dental checkup this year?

    Answer: 0.746 or 0.7462 Type: S   Sect: 4 Obj: 4   RANDOM: Y

79. The probability that a person will have 1 dental checkup per year is 0.5. If 9 people
    are selected at random, what is the probability that at most 5 will have 1 dental
    checkup this year?

    Answer: 0.7462 or 0.746 Type: S   Sect: 4 Obj: 4   RANDOM: Y

80. The probability that a person will have 1 dental checkup per year is 0.5. If 9 people are selected at random, what is the probability that exactly 7 will have 1 dental checkup this year?

    Answer: 0.0703 Type: S   Sect: 4 Obj: 4   RANDOM: Y

81. If the probability that a person will have 2 dental checkups a year is 0.3, in a group of 1,500 people, how many would you expect to have 2 dental checkups this year?

    Answer: 450 Type: S   Sect: 5 Obj: 5   RANDOM: Y

82. Find the mean for the number of kings drawn from a standard deck, with replacement, if one card is drawn 75 times.

    Answer: 5.7692 Type: S   Sect: 5 Obj: 5   RANDOM: Y

83. In the town of Rawlins, it was estimated that 7 out of 15 people ate dinner at a restaurant on Saturday night. If there are 12,000 people in Rawlins, how many would you expect to be dining out on Saturday night?

    Answer: 5,600 Type: S   Sect: 5 Obj: 5   RANDOM: Y

84. If 2% of the bolts made by an airplane factory are defective, how many good bolts would you expect to find in a shipment of 25,000?

    Answer: 24,500 Type: S   Sect: 5 Obj: 5   RANDOM: Y

85. It has been estimated that 39% of Americans own a dog. What are the mean and standard deviation for the number of Americans who own a dog in a sample of 50,000?

    Answer: mean = 19,500, standard deviation = 109.0642 Type: S   Sect: 5 Obj: 5

    RANDOM: Y

86. A salesman of large appliances has determined that the probability that a customer looking at his products will actually buy one is 0.02. What is the probability that out of the next 25 customers, exactly 3 will buy an appliance?

    Answer: 0.0118 Type: S   Sect: 3 Obj: 4   RANDOM: Y

87. A salesman of large appliances has determined that the probability that a customer looking at his products will actually buy one is 0.02. What is the probability that out of the next 25 customers, none will buy?

    Answer: 0.6035 Type: S   Sect: 5 Obj: 4   RANDOM: Y

88. A salesman of large appliances has determined that the probability that a customer looking at his products will actually buy one is 0.04. What is the probability that out of the next 30 customers, at least one person will buy his product?

Answer: 0.7061 Type: S   Sect: 3 Obj: 4   RANDOM: Y

89. A survey showed that 64% of Americans live in the state in which they were born. What is the probability that in the next group of 10 people, exactly 7 live in the state in which they were born?

Answer: 0.2462 Type: S   Sect: 3 Obj: 4   RANDOM: Y

90. A survey showed that 64% of Americans live in the state in which they were born. What is the probability that in the next group of 10 people, exactly 5 live in the state in which they were born?

Answer: 0.1636 Type: S   Sect: 3 Obj: 4   RANDOM: Y

91. A survey showed that 65% of Americans live in the state in which they were born. In a group of 20 people, what is the probability that exactly 9 do not live in the state in which they were born?

Answer: 0.1158 Type: S   Sect: 4 Obj: 4   RANDOM: Y

TRUE/FALSE

92. The number of ants making an anthill is an example of a discrete random variable.

Answer: T   Type: T   Sect: 1 Obj: 6   RANDOM: Y

93. The time it takes a broken leg to heal is an example of a discrete random variable.

Answer: F   Type: T   Sect: 1 Obj: 6   RANDOM: Y

94. Another name for the expected value of a random variable is the mode.

Answer: F   Type: T   Sect: 1 Obj: 6   RANDOM: Y

95. A fair game is one with an expected value of 0 for all players.

Answer: T   Type: T   Sect: 2 Obj: 6   RANDOM: Y

MULTIPLE CHOICE

96. If X represents the number of jurors on a 9-member panel who vote for an acquittal, which of the following values cannot be a value for X?
   a. 0   b. 5   c. 12   d. 9

   Answer: c Type: M   Sect: 2 Obj: 6   RANDOM: Y

97. The probability associated with a failure is:

   a. p b. $1 - p$ c. $\overline{X}$ d. n

   Answer: b Type: M   Sect: 3 Obj: 6   RANDOM: Y

98. If there are $n = 35$ trials and $X = 20$ successes, how many failures are there?
   a. 26   b. 35   c. 15   d. 55

   Answer: c Type: M   Sect: 3 Obj: 6   RANDOM: Y

99. How many ways can there be 4 successes in 6 trials?
   a. 15   b. 2.5   c. 24   d. 30

   Answer: a Type: M   Sect: 3 Obj: 6   RANDOM: Y

100. How many ways can there be 3 successes in 8 trials?
   a. 6,720   b. 336   c. 56   d. 24

   Answer: c Type: M   Sect: 3 Obj: 6   RANDOM: Y

101. How many ways can there be no successes in 6 trials?
   a. 6   b. 0   c. 1   d. 12

   Answer: c Type: M   Sect: 3 Obj: 6   RANDOM: Y

SHORT ANSWER

102. A card is drawn from a standard deck and then replaced. Let X be the number of hearts drawn in 20 draws. What is p?

   Answer: 1/4 or .25 Type: S   Sect: 3 Obj: 6   RANDOM: Y

103. When we construct a histogram of the probability distribution of a binomial random variable, how do we label the vertical axis?

   Answer: probability Type: S   Sect: 3 Obj: 6   RANDOM: Y

104. In a histogram for a probability distribution, what is the sum of the areas of all the bars?

Answer: one Type: S   Sect: 5 Obj: 6   RANDOM: Y

MULTIPLE CHOICE

****
The remaining questions in this chapter require the use of the formula generation capabilities of ESATEST III. Please see the ESATEST III documentation if you need directions on how to use them.
****

105. A binomial distribution has n = @5, p = @0.35, and x = `@2`. What is the probability associated with this data?
    a. I 0.3124
    b. I 0.3364
    c. I 0.1811
    d. I 0.3456

Answer: @1 c@3*@2^@3*(1-@2)^(@1-@3),5,5,20,0,0.35,0.2,0.7,2,2,1,5,0,[4]   Type: M

RANDOM: F

106. If we have a binomial distribution with n = @200 and p = @0.47, what would be the variance?
    a. I 49.82
    b. I 94
    c. I 7.06
    d. I 9.70

Answer: @1*@2*(1-@2),200,100,300,0,0.47,0.3,0.8,2,[2]   Type: M   RANDOM: F

SHORT ANSWER

107. If n = @6, p = 2/@3, and x = @4, what is the probability?

Answer: @1 c@3*(2/@2)^@3*(1-2/@2)^(@1-@3),6,6,20,0,3,3,9,(3,5,7,9),4,1,5,0,[3]   Type: S

RANDOM: F

CHAPTER 6

TRUE/FALSE

1. In a normal distribution the mean, median, and mode are located at the center.

   Answer: T Type: T   Sect: 2 Obj: 1   RANDOM: Y

2. The area under the normal distribution curve between $z = 1.52$ and $z = 2.35$ would be 0.9263.

   Answer: F Type: T   Sect: 3 Obj: 2   RANDOM: Y

3. The area to the left of $z = -1.47$ would be 0.0708.

   Answer: T Type: T   Sect: 3 Obj: 2   RANDOM: Y

4. $P(z > 1.25)$ would be 0.3944.

   Answer: F Type: T   Sect: 3 Obj: 2   RANDOM: Y

5. The z-value such that the area under the normal distribution to the right of it would be 0.0049 is a value of 2.58.

   Answer: T Type: T   Sect: 3 Obj: 4   RANDOM: Y

6. The average number of days a person will be sick this year is 7 with a standard deviation of 2. If a person is picked at random, the probability that he or she will be sick less than 5 days would be 0.3413.

   Answer: F Type: T   Sect: 5 Obj: 5   RANDOM: Y

7. A manager wants to fire the bottom 6% of his crew. The z-score that would be used to calculate the cutoff point is 1.47.

   Answer: F Type: T   Sect: 3 Obj: 6   RANDOM: Y

8. A manager wants to give a bonus to the top 3% of his salesclerks. The z-score which would be used to calculate the cutoff point would be 1.88.

   Answer: T   Type: T Sect: 3 Obj: 6   RANDOM: Y

9. $P(z < 2.32)$ would be 0.4898.

   Answer: F   Type: T Sect: 3 Obj: 2   RANDOM: Y

10. The brick company found that 4% of its bricks were defective. If 400 bricks are checked at random, the probability that there are more than 18 defective bricks could be found by using the normal approximation method.

    Answer: T Type: T   Sect: 7 Obj: 9   RANDOM: Y

MULTIPLE CHOICE

11. The area under the normal curve that lies within two standard deviations of the mean is approximately what percent?
    a. 68%
    b. 99.7%
    c. 75%
    d. 95%

    Answer: d Type: M   Sect: 3 Obj: 1   RANDOM: Y

12. What is the area under the normal curve between $z = -1.92$ and $z = -1.21$?
    a. 0.7400
    b. 0.0857
    c. 0.8595
    d. 0.1405

    Answer: b Type: M   Sect: 3 Obj: 2   RANDOM: Y

13. What is the area to the left of $z = 2.03$?
    a. 0.9788
    b. 0.4788
    c. 0.0212
    d. 0.9576

    Answer: a Type: M   Sect: 3 Obj: 2   RANDOM: Y

14. Using the standard normal distribution, what is $P(-0.25 < z < 1.35)$?
    a. 0.4937
    b. 0.3128
    c. 0.5102
    d. 0.3643

    Answer: c Type: M   Sect: 3 Obj: 3   RANDOM: Y

15. The number of ties sold daily at a men's clothing store was approximately normally distributed with a mean of 15 and a standard deviation of 3. If a day is picked at random, what is the probability that the store will sell over 20 ties?
    a. 0.0475
    b. 0.4525
    c. 0.0485
    d. 0.4515

    Answer: a Type: M   Sect: 5 Obj: 5   RANDOM: Y

16. What is the z-value corresponding to the 35th percentile?
    a. -1.04
    b. -0.39
    c. -0.35
    d. -0.65

    Answer: b Type: M   Sect: 3 Obj: 4   RANDOM: Y

17. The average age of farm workers in the United States was thought to be 45 with a standard deviation of 7 years. If a farmer is selected at random, what is the probability that the farmer is over 55? (Assume a normal distribution.)
    a. 0.4236   b. 0.0764   c. 0.9236   d. 0.0778

    Answer: b Type: M   Sect: 5 Obj: 5   RANDOM: Y

18. The average age of farm workers in the United States was thought to be 45 with a standard deviation of 7 years. If a farmer is selected at random, what is the probability that he or she is between 42 and 53 years old? (Assume a normal distribution.)
    a. 0.5393   b. 0.2611   c. 0.2065   d. 0.5357

    Answer: a Type: M   Sect: 5 Obj: 5   RANDOM: Y

19. A brick company found that 4% of its bricks were defective. If 400 bricks are checked at random, what is the probability that fewer than 15 are defective?
    a. 0.1480   b. 0.4483   c. 0.352   d. 0.3974

    Answer: c Type: M   Sect: 7 Obj: 7   RANDOM: Y

20. A brick company found that 4% of its bricks were defective. If 400 bricks are checked at random, what is the probability that there are more than 18 defective bricks?
    a. 0.2389
    b. 0.2611
    c. 0.4364
    d. 0.3520

    Answer: b Type: M   Sect: 7 Obj: 7   RANDOM: Y

SHORT ANSWER

Find the area under the normal distribution for the given z-values.

21. What is the area between $z = 0$ and $z = -1.52$?

   Answer: 0.4357 Type: S   Sect: 3 Obj: 2   RANDOM: Y

22. What is the area between $z = 0$ and $z = 2.43$?

   Answer: 0.4925 Type: S   Sect: 3 Obj: 2   RANDOM: Y

23. What is the area between $z = -1.34$ and $z = 0.31$?

   Answer: 0.5316 Type: S   Sect: 3 Obj: 2   RANDOM: Y

24. What is the area between $z = -2.05$ and $z = -0.87$?

   Answer: 0.172 Type: S   Sect: 3 Obj: 2   RANDOM: Y

25. What is the area between $z = 0.76$ and $z = 0.89$?

   Answer: 0.0369 Type: S   Sect: 3 Obj: 2   RANDOM: Y

26. What is the area to the right of $z = -0.29$?

   Answer: 0.3859 Type: S   Sect: 3 Obj: 2   RANDOM: Y

27. What is the area to the right of $z = 2.89$?

   Answer: 0.0019 Type: S   Sect: 3 Obj: 2   RANDOM: Y

28. What is the area to the left of $z = 1.85$?

   Answer: 0.9678 Type: S   Sect: 3 Obj: 2   RANDOM: Y

29. What is the area to the left of $z = -1.07$?

   Answer: 0.1423 Type: S   Sect: 3 Obj: 2   RANDOM: Y

30. What is the area to the right of $z = 1.96$ and to the left at $z = -2.56$?

   Answer: 0.0302 Type: S   Sect: 3 Obj: 2   RANDOM: Y

Find the probability, using the standard normal.

31. $P(z > 1.23)$

    Answer: 0.1093 Type: S   Sect: 3 Obj: 3   RANDOM: Y

32. $P(0 < z < 1.1)$

    Answer: 0.3643 Type: S   Sect: 3 Obj: 3   RANDOM: Y

33. $P(-0.36 < z \leq 0)$

    Answer: 0.1406 Type: S   Sect: 3 Obj: 3   RANDOM: Y

34. $P(-1.35 \leq z < 0.23)$

    Answer: 0.5025 Type: S   Sect: 3 Obj: 3   RANDOM: Y

35. $P(2.54 < z < 2.98)$

    Answer: 0.0041 Type: S   Sect: 3 Obj: 3   RANDOM: Y

36. $P(z > -1.84)$

    Answer: 0.9671 Type: S   Sect: 3 Obj: 3   RANDOM: Y

37. $P(z < 1.39)$

    Answer: 0.9177 Type: S   Sect: 3 Obj: 3   RANDOM: Y

38. $P(z \geq 0.12)$

    Answer: 0.4522 Type: S   Sect: 3 Obj: 3   RANDOM: Y

39. $P(z \leq -0.4)$

    Answer: 0.3446 Type: S   Sect: 3 Obj: 3   RANDOM: Y

40. $P(-0.57 \leq z < -0.03)$

    Answer: 0.2037 Type: S   Sect: 3 Obj: 3   RANDOM: Y

Find the z-value.

41. What is the z-value that corresponds to the 30th percentile?

    Answer: -0.52 Type: S   Sect: 3 Obj: 4   RANDOM: Y

42. What is the z-value such that 42% of the total area lies to the left of it?

    Answer: -0.20 Type: S   Sect: 3 Obj: 4   RANDOM: Y

43. What is the z-value to the right of the mean such that 87.49% of the distribution lies to the left of it?

    Answer: 1.15 Type: S   Sect: 3 Obj: 4   RANDOM: Y

44. What is the z-value to the left of the mean such that 98.81% of the distribution lies to the right of it?

    Answer: -2.26 Type: S   Sect: 3 Obj: 4   RANDOM: Y

45. What is the z-value for the 99th percentile?

    Answer: 2.33 Type: S   Sect: 3 Obj: 4   RANDOM: Y

46. What is the z-value for the 29th percentile?

    Answer: -0.55 Type: S   Sect: 3 Obj: 4   RANDOM: Y

Directions for the problems below: Assume that the following distribution is a normal probability distribution.

47. If the average weight of a box of cereal is 20 ounces with a standard deviation of 0.15 ounce, what is the probability that a box of cereal will weigh more than 20.4 ounces?

    Answer: 0.0038 Type: S   Sect: 5 Obj: 5   RANDOM: Y

48. If the average weight of a box of cereal is 20 ounces with a standard deviation of 0.15 ounce, what is the probability that a box of cereal will weigh less than 19.8 ounces?

    Answer: 0.0918 Type: S   Sect: 5 Obj: 5   RANDOM: Y

49. If the average weight of a box of cereal is 20 ounces with a standard deviation of 0.15 ounce, what is the probability that a box of cereal will weigh between 19.7 and 20.3 ounces?

    Answer: 0.9544 Type: S   Sect: 5 Obj: 5   RANDOM: Y

50. A company claims its new car gets 45 miles per gallon with a standard deviation of 3 miles per gallon. What is the probability that the car will get more than 50 miles per gallon?

    Answer: 0.0475 Type: S   Sect: 5 Obj: 5   RANDOM: Y

51. If the average college student studies 20 hours a week, with a standard deviation of 2.5 hours, what is the probability that a student will study more than 27 hours?

    Answer: 0.0026 Type: S   Sect: 5 Obj: 5   RANDOM: Y

52. If the average college student studies 20 hours a week, with a standard deviation of 2.5 hours, what is the probability that a student will study between 17 and 22 hours?

    Answer: 0.6730 Type: S   Sect: 5 Obj: 5   RANDOM: Y

53. If the average college student studies 20 hours a week, with a standard deviation of 2.5 hours, what is the probability that a student will study less than 16 hours?

    Answer: 0.0548 Type: S   Sect: 5 Obj: 5   RANDOM: Y

54. A study showed that the average college student studied 20 hours a week, with a standard deviation of 2.5 hours. If Culver College had 4,000 students, how many would you expect to find who studied more than 25 hours?

    Answer: approximately 91 Type: S   Sect: 5 Obj: 5   RANDOM: Y

55. The average number of words per minute that a reading class can read is 175, with a standard deviation of 20 words per minute. What is the probability that a randomly selected person can read more than 240 words per minute?

    Answer: 0.0006 Type: S   Sect: 5 Obj: 5   RANDOM: Y

56. The average number of words per minute that a reading class can read is 175, with a standard deviation of 20 words per minute. What is the probability that a randomly selected person can read less than 150 words per minute?

    Answer: 0.1056 Type: S   Sect: 5 Obj: 5   RANDOM: Y

57. The average number of words per minute that a reading class can read is 175, with a standard deviation of 20 words per minute. If there were 60 people in the class, how many people would you expect to be able to read between 170 and 182 words per minute?

    Answer: approximately 14 Type: S   Sect: 5 Obj: 5   RANDOM: Y

58. The average number of words per minute that a reading class can read is 175, with a standard deviation of 20 words per minute. The top 3% of the class is to receive a special award. How many words per minute would a person have to read in order to get an award?

    Answer: 213 or more Type: S   Sect: 6 Obj: 6   RANDOM: Y

59. In a normal distribution, what is the mean when the standard deviation is 18 and 2.6% of the distribution lies to the right of 72?

    Answer: $\mu = 37.08$ Type: S   Sect: 6 Obj: 6   RANDOM: Y

60. In a normal distribution, what is the standard deviation when $\mu = 127$ and 7.5% of the area lies to the left of 97?

    Answer: standard deviation is 20.83 Type: S   Sect: 6 Obj: 6   RANDOM: Y

61. If a normal distribution has a mean of 65 and a standard deviation of 10, what is the value of the 67th percentile?

    Answer: 69.4 Type: S   Sect: 6 Obj: 6   RANDOM: Y

62. If a normal distribution has a mean of 126 and a standard deviation of 22, what is the value of the 34th percentile?

    Answer: 116.98 Type: S   Sect: 6 Obj: 6   RANDOM: Y

63. If a normal distribution has a mean of 300 and a standard deviation of 16, what is the value of the 58th percentile?

    Answer: 303.2 Type: S   Sect: 6 Obj: 6   RANDOM: Y

64. If a normal distribution has a mean of 600 and a standard deviation of 25, what is the value of the 98th percentile?

    Answer: 651.25 Type: S   Sect: 6 Obj: 6   RANDOM: Y

65. In a normal distribution, what is the mean when the standard deviation is 20 and 3.8% of the distribution lies to the left of 200?

    Answer: 235.4 Type: S   Sect: 6 Obj: 6   RANDOM: Y

66. In a normal distribution, what is the mean when the standard deviation is 30 and 18.2% of the distribution lies to the right of 350?

    Answer: 322.7 Type: S   Sect: 6 Obj: 6   RANDOM: Y

67. The weights of 10-year-old girls are normally distributed with a mean of 80 pounds and a standard deviation of 6 pounds. A girl who weighs at the 97th percentile is considered overweight. What would a 10-year-old girl have to weigh in order to be considered overweight?

    Answer: 91.3 pounds Type: S   Sect: 6 Obj: 6   RANDOM: Y

68. The weights of 10-year-old girls are normally distributed with a mean of 80 pounds and a standard deviation of 6 pounds. What is the probability that a girl selected at random would weigh less than 70 pounds?

    Answer: 0.0475 Type: S   Sect: 5 Obj: 5   RANDOM: Y

69. The weights of 10-year-old girls are normally distributed with a mean of 80 pounds and a standard deviation of 6 pounds. What is the probability that a 10-year-old girl selected at random would weigh more than 88 pounds?

    Answer: 0.0918 Type: S   Sect: 5 Obj: 5   RANDOM: Y

70. The weights of 10-year-old girls are normally distributed with a mean of 80 pounds and a standard deviation of 6 pounds.   What is the probability that a 10-year-old girl selected at random would weigh between 77 and 85 pounds?

    Answer: 0.4882 Type: S   Sect: 5 Obj: 5   RANDOM: Y

71. The grade-point averages of the students at Alpha Beta fraternity are normally distributed with a mean of 2.75 and a standard deviation of 0.2. What is the probability that a randomly selected member of Alpha Beta will have a grade-point average of more than 3.3?

    Answer: 0.003 Type: S   Sect: 5 Obj: 5   RANDOM: Y

72. The grade-point averages of the members of Alpha Beta fraternity are normally distributed with a mean of 2.75 and a standard deviation of 0.2. What is the probability that a randomly selected member of Alpha Beta will have a grade-point average of less than 2.6?

    Answer: 0.2266 Type: S   Sect: 5 Obj: 5   RANDOM: Y

73. The grade-point averages of the students at Alpha Beta fraternity are normally distributed with a mean of 2.75 and a standard deviation of 0.2. What is the probability that a randomly selected member of Alpha Beta will have a grade-point average between 2.5 and 3.1?

    Answer: 0.8543 Type: S   Sect: 5 Obj: 5   RANDOM: Y

74. The grade-point averages of the students at Alpha Beta fraternity are normally distributed with a mean of 2.75 and a standard deviation of 0.2. If the top 10% of the students are to get a free dinner, what would be the lowest grade-point average a student could have and get the free dinner?

    Answer: about 3.01 Type: S   Sect: 5 Obj: 6   RANDOM: Y

75. One type of children's TV dinner has calories that are normally distributed with a mean of 350 calories and a standard deviation of 20 calories. If a TV dinner is selected at random, what is the probability that it contains more than 400 calories?

    Answer: 0.0062 Type: S   Sect: 5 Obj: 5   RANDOM: Y

76. One type of children's TV dinner has calories that are normally distributed with a mean of 350 calories and a standard deviation of 20 calories. If a TV dinner is selected at random, what is the probability that it contains less than 340 calories?

    Answer: 0.3085 Type: S   Sect: 5 Obj: 5   RANDOM: Y

77. One type of children's TV dinner has calories that are normally distributed with a mean of 350 calories and a standard deviation of 20 calories. If a TV dinner is selected at random, what is the probability that it contains between 335 and 348 calories?

    Answer: 0.2336 Type: S   Sect: 5 Obj: 5   RANDOM: Y

78. One type of children's TV dinner has calories that are normally distributed with a mean of 350 calories and a standard deviation of 20 calories. If a TV dinner is selected at random, what is the probability that it contains between 350 and 378 calories?

    Answer: 0.4192 Type: S   Sect: 5 Obj: 5   RANDOM: Y

79.A survey of lobster fishermen found that they catch an average of 32 pounds of lobsters a day with a standard deviation of 4 pounds. Assuming a normal distribution, what is the probability that a randomly selected fisherman will catch 42 pounds of lobsters today?

Answer: 0.0062 Type: S   Sect: 5 Obj: 5   RANDOM: Y

80.A survey of lobster fishermen found that they catch an average of 32 pounds of lobsters a day with a variance of 16. Assuming a normal distribution, what is the probability that a randomly selected fisherman will catch between 31 and 40 pounds of lobsters?

Answer: 0.5759 Type: S   Sect: 5 Obj: 5   RANDOM: Y

81.A survey of lobster fishermen found that they catch an average of 32 pounds of lobsters a day with a standard deviation of 4 pounds. Assume a normal distribution. The bottom 1% of lobster fishermen will receive a booby prize. How many pounds of fish would you have to catch to receive the booby prize?

Answer: 22.7 lbs. or less Type: S   Sect: 6 Obj: 6   RANDOM: Y

82.In a recent study, 44% of grocery shoppers said they were eating more seafood. If 500 shoppers were surveyed, what is the probability that at least 240 were eating more seafood?

Answer: 0.0392 Type: S   Sect: 7 Obj: 7   RANDOM: Y

83.In a recent study, 44% of grocery shoppers said they were eating more seafood. If 500 shoppers were surveyed, what is the probability that fewer than 215 were eating more seafood?

Answer: 0.3085 Type: S   Sect: 7 Obj: 7   RANDOM: Y

84.In a recent study, 44% of grocery shoppers said they were eating more seafood. If 500 shoppers were surveyed, what is the probability that between 214 and 234 shoppers were eating more seafood?

Answer: 0.5803 Type: S   Sect: 7 Obj: 7   RANDOM: Y

85.If Mary averages 82% on her quizzes, what is the probability that she will get at least 160 points out of 200 on her final?

Answer: 0.7967 Type: S   Sect: 6 Obj: 7   RANDOM: Y

86.If Mary averages 82% on her quizzes, what is the probability that she will get more than 170 points out of 200 on her final?

Answer: 0.1151 Type: S   Sect: 6 Obj: 7   RANDOM: Y

87. A national poll claims that 75% of all Americans do not want higher taxes. If 1,600 people were polled, what is the probability that exactly 1,195 people did not want higher taxes?

    Answer: 0.0229 Type: S   Sect: 6 Obj: 7   RANDOM: Y

88. A national poll claims that 75% of all Americans do not want higher taxes. If 1,600 people were polled, what is the probability that at most 1,240 people did not want higher taxes?

    Answer: 0.9904 Type: S   Sect: 6 Obj: 7   RANDOM: Y

89. A national poll claims that 75% of all Americans do not want higher taxes. If 1,600 people were polled, what is the probability that more than 1,185 people did not want higher taxes?

    Answer: 0.7995 Type: S   Sect: 6 Obj: 7   RANDOM: Y

90. If the probability that a newborn child will be a male is 48%, what is the probability that in a random sample of 40 families exactly 17 will have boys?

    Answer: 0.0969 Type: S   Sect: 6 Obj: 7   RANDOM: Y

91. If the probability that a newborn child will be a male is 48%, what is the probability that in a random sample of 40 families more than 15 will have boys?

    Answer: 0.879 Type: S   Sect: 6 Obj: 7   RANDOM: Y

92. If the probability that a newborn child will be a male is 48%, what is the probability that in a random sample of 40 families fewer than 18 will have boys?

    Answer: 0.2946 Type: S   Sect: 6 Obj: 7   RANDOM: Y

Check the binomial to see if it can be approximated by using the normal. If it can't, tell why not.

93. $n = 18,\ p = 0.2$

    Answer: Cannot use the normal approximation method. $np < 5$ Type: S   Sect: 6

    Obj: 7   RANDOM: Y

94. $n = 50,\ p = 0.05$

    Answer: Cannot use the normal approximation method. $np < 5$ Type: S   Sect: 6

    Obj: 7   RANDOM: Y

95. $n = 40$, $p = 0.3$

   Answer: Can use the normal approximation method.   Type: S   Sect: 6   Obj: 7

   RANDOM: Y

96. $n = 30$, $p = 0.8$

   Answer: Can use the normal approximation method.   Type: S   Sect: 6   Obj: 7

   RANDOM: Y

MULTIPLE CHOICE

****
The remaining questions in this chapter require the use of the formula
generation capabilities of ESATEST III.   Please see the ESATEST III
documentation if you need directions on how to use them.
****

97. What is the area under the normal curve between $z = @-1.92$ and $z = @-1.21$?
    a. l 0.7400
    b. l 0.0857
    c. l 0.8595
    d. l 0.1405

   Answer: snd(@1)-snd(@2),-1.92,-2,-1.5,2,-1.21,-1.5,-1,2,[4]   Type: M   RANDOM: F

98. Using the standard normal distribution, what is $P(@-0.25 < z < @1.35)$?
    a. l 0.4937
    b. l 0.3128
    c. l 0.5102
    d. l 0.3643

   Answer: snd(@1)+snd(@2),-0.25,-0.2,-0.5,2,1.35,1,2,2,[4]   Type: M   RANDOM: F

SHORT ANSWER

Find the area under the normal distribution for the given z-values.

99. What is the area between $z = 0$ and $z = @-1.52$?

   Answer: snd(@1),-1.52,-2,0,2,[4]   Type: S   RANDOM: F

100. What is the area between $z = @-1.34$ and $z = @0.31$?

     Answer: snd(@1)+snd(@2),-1.34,-2,-1,2,0.31,0,2,2,[4]   Type: S   RANDOM: F

101. What is the area between $z = @0.76$ and $z = @0.89$?

     Answer: snd(@2)-snd(@1),0.76,0.5,0.8,2,0.89,0.8,2,2,[4]   Type: S   RANDOM: F

102. What is the area to the right of $z = @-0.29$?

     Answer: 0.5+snd(@1),-0.29,-2,-1,2,[4]   Type: S   RANDOM: F

103. What is the area to the right of $z = @2.89$?

     Answer: 0.5-snd(@1),2.89,1,3,2,[4]   Type: S   RANDOM: F

104. What is the area to the left of $z = @-1.07$?

     Answer: 0.5-snd(@1),-1.07,-2,-1,2,[4]   Type: S   RANDOM: F

105. What is the area to the right of $z = @1.96$ and to the left of $z = @-2.56$?

     Answer: 1-snd(@1)-snd(@2),1.96,1,2,2,-2.56,-3,-1,2,[4]   Type: S   RANDOM: F

106. Find the probability, using the standard normal. $P(z > @1.23)$

     Answer: 0.5-snd(@1),1.23,1,3,2,[4]   Type: S   RANDOM: F

107. Find the probability, using the standard normal. $P(0 < z < @1.1)$

     Answer: snd(@1),1.1,1,3,2,[4]   Type: S   RANDOM: F

108. Find the probability, using the standard normal. $P(@-0.36 < z \leq 0)$

     Answer: snd(@1),-0.36,-3,0,2,[4]   Type: S   RANDOM: F

109. Find the probability, using the standard normal. $P(@-1.35 \leq z < @0.23)$

     Answer: snd(@1)+snd(@2),-1.35,-2,-1,2,0.23,0,2,2,[4]   Type: S   RANDOM: F

110. Find the probability, using the standard normal. $P(@2.54 < z < @2.98)$

     Answer: snd(@2)-snd(@1),2.54,2,2.5,2,2.98,2.5,3,2,[4]   Type: S   RANDOM: F

111. Find the probability, using the standard normal. $P(z > @-1.84)$

   Answer: 0.5+snd(@1),-1.84,-2,0,2,[4] Type: S   RANDOM: F

112. Find the probability, using the standard normal. $P(z < @1.39)$

   Answer: 0.5+snd(@1),1.39,1,3,2,[4] Type: S   RANDOM: F

113. Find the probability, using the standard normal. $P(z \geq @0.12)$

   Answer: 0.5-snd(@1),0.12,0,3,2,[4] Type: S   RANDOM: F

114. Find the probability, using the standard normal. $P(z \leq @-0.4)$

   Answer: 0.5-snd(@1),-0.4,-3,0,2,[4] Type: S   RANDOM: F

115. Find the probability, using the standard normal.
   $P(@-0.57 \leq z < @-0.03)$

   Answer: snd(@1)-snd(@2),-0.57,-2,-0.5,2,-0.03,-0.5,0,2,[4] Type: S   RANDOM: F

MULTIPLE CHOICE

116. What is the area to the left of $z = @2.03$?
   a. | 0.9788
   b. | 0.4788
   c. | 0.0212
   d. | 0.9576

   Answer: 0.5+snd(@1),2,0,3,2,[4] Type: M   RANDOM: F

CHAPTER 7

TRUE/FALSE

1. Population characteristics or values are called parameters.

   Answer: T Type: T Sect: 1 Obj: 1   RANDOM: Y

2. The inferential process consists of using parameters to estimate statistics.

   Answer: F   Type: T Sect: 1 Obj: 1   RANDOM: Y

3. In a stratified sample, the proportion of each segment in the sample would correspond to the proportion of each segment in the population.

   Answer: T Type: T Sect: 1 Obj: 1   RANDOM: Y

4. A sampling distribution depends only on the particular sample size, n, being investigated.

   Answer: F Type: T Sect: 2 Obj: 1   RANDOM: Y

5. In a random sampling process, each of the possible samples is equally likely to occur.

   Answer: T Type: T Sect: 2 Obj: 1   RANDOM: Y

6. Constructing a sampling distribution requires finding all possible values of the statistic and probability associated with each of these values.

   Answer: T Type: T Sect: 2 Obj: 1   RANDOM: Y

7. The mean of the sample means is equal to the standard deviation divided by the sample size.

   Answer: F Type: T Sect: 3 Obj: 2   RANDOM: Y

8. If the standard deviation of a population is 35 and $n = 49$, the standard error of the mean is 5.

   Answer: T Type: T Sect: 4 Obj: 2   RANDOM: Y

9. If the standard deviation of a large population is 100 and $n = 400$, the standard error of the mean is 0.25.

   Answer: F Type: T Sect: 4 Obj: 2   RANDOM: Y

10. If the variance of a large population is 100 and $n = 64$, the standard error of the mean is 12.5.

    Answer: F Type: T   Sect: 4 Obj: 2   RANDOM: Y

11. The finite correction factor when $N = 500$ and $n = 70$ is 0.86.

    Answer: F Type: T   Sect: 7 Obj: 3   RANDOM: Y

12. The finite correction factor when $N = 1000$ and $n = 125$ is approximately 0.936.

    Answer: T Type: T   Sect: 7 Obj: 3   RANDOM: Y

13. If $N = 600$ and $n = 150$, the standard error of the mean will have to be computed using the finite population correction factor.

    Answer: T Type: T   Sect: 7 Obj: 3   RANDOM: Y

14. If $N = 1,500$ and $n = 50$, the finite population correction factor will need to be calculated in order to compute the standard error of the mean.

    Answer: F Type: T   Sect: 7 Obj: 3   RANDOM: Y

15. The central limit theorem states that the mean of the sample means will be the same as the population mean.

    Answer: T Type: T   Sect: 5 Obj: 2   RANDOM: Y

16. If $N = 5,000$ and $n = 300$, the finite correction factor would be used.

    Answer: T Type: T   Sect: 7 Obj: 3   RANDOM: Y

MULTIPLE CHOICE

17. A population has a mean of 30 and a standard deviation of 60. A sample of 36 items has a mean of 33. What is the standard error of the mean for these data?
    a. 5
    b. 10
    c. 1.67
    d. 0.83

    Answer: b Type: M   Sect: 4 Obj: 2   RANDOM: Y

18. The average age of farm workers in the United States was thought to be 45 with a standard deviation of 7 years. If 144 farm workers are selected at random, what is the probability that their mean age is more than 46?
    a. 0.4564
    b. 0.4554
    c. 0.0446
    d. 0.0436

    Answer: d Type: M   Sect: 5 Obj: 2   RANDOM: Y

19. The standard deviation of a population is 50, $N = 600$ and $n = 50$. Using the correction factor, what is the adjusted standard error of the mean?
    a. 6.776
    b. 7.071
    c. 4.867
    d. 6.493

    Answer: a Type: M   Sect: 7 Obj: 3   RANDOM: Y

20. What are sample values, such as $\overline{X}$, called?
    a. statistics   b. parameters
    c. inferences   d. data

    Answer: a Type: M   Sect: 1 Obj: 1   RANDOM: Y

21. An administrator wants to select a random sample of 25 students from a campus with 4,530 students. Which of the following values from a random number table is not usable?
    a. 152   b. 4,175   c. 1,567   d. 4,621

    Answer: d Type: M   Sect: 1 Obj: 1   RANDOM: Y

22. Use the following sampling distribution to answer the question.

$\overline{X}$	$P(\overline{X})$
3	1/11
5	2/11
7	5/11
10	1/11
15	2/11

    What is $P(6 < X \le 10)$?
    a. 8/11   b. 5/11   c. 6/11   d. 3/11

    Answer: c Type: M   Sect: 2 Obj: 1   RANDOM: Y

23.Use the following sampling distribution to answer the question.

$\overline{X}$	$P(\overline{X})$
3	1/11
5	2/11
7	5/11
10	1/11
15	2/11

What is $P(X > 7)$?
a. 5/11 b. 8/11 c. 7/11 d. 3/11

Answer: d Type: M Sect: 2 Obj: 1 RANDOM: Y

24.Use the following sampling distribution to answer the question.

$\overline{X}$	$P(\overline{X})$
3	1/11
5	2/11
7	5/11
10	1/11
15	2/11

What is $P(X \leq 10)$?
a. 1/11 b. 9/11 c. 8/11 d. 7/11

Answer: b Type: M Sect: 2 Obj: 1 RANDOM: Y

25.The mean of a population is 400 and the standard deviation is 50. A sample of size 225 is selected. What is the standard error of the mean?
a. 0.22 b. 3.33 c. 1.78 d. 8

Answer: b Type: M Sect: 4 Obj: 2 RANDOM: Y

26.The mean of a population is 120 and the standard deviation is 40. A sample of size 100 is selected. What is the standard error of the mean?
a. 4 b. 0.4 c. 12 d. 1.2

Answer: a Type: M Sect: 4 Obj: 2 RANDOM: Y

27.The mean of a population is 800 and the variance is 64. A sample of size 400 is selected. What is the standard error of the mean?
a. 0.16 b. 3.2 c. 0.4 d. 2.0

Answer: c Type: M Sect: 4 Obj: 2 RANDOM: Y

28. The standard deviation of a population is 15, $N = 100$, and $n = 25$. Using the finite population correction factor, what is the adjusted standard error of the mean?
    a. 2.61   b. 2.27   c. 1.14   d. 0.52

    Answer: a Type: M   Sect: 7 Obj: 3   RANDOM: Y

29. If $N = 1,000$ and $n = 150$, what is the finite population correction factor?
    a. 0.85   b. 0.92   c. 0.72   d. 0.87

    Answer: b Type: M   Sect: 7 Obj: 3   RANDOM: Y

30. If $N = 20,000$ and $n = 1,800$, what is the finite population correction factor?
    a. 0.95   b. 0.91   c. 0.83   d. 0.98

    Answer: a Type: M   Sect: 7 Obj: 3   RANDOM: Y

31. If the finite correction factor is 0.92 and $N = 2,500$, what is the value of n?
    a. 103   b. 385   c. 1375   d. 535

    Answer: b Type: M   Sect: 7 Obj: 3   RANDOM: Y

32. If the finite population correction factor is 0.87 and $N = 750$, what is the value of n?
    a. 147   b. 51   c. 183   d. 119

    Answer: c Type: M   Sect: 7 Obj: 3   RANDOM: Y

SHORT ANSWER

33. Using the following sampling distribution, what is the mean of this distribution?

$\overline{X}$	$P(\overline{X})$
3	1/11
5	2/11
7	5/11
10	1/11
15	2/11

    Answer: 8 Type: S   Sect: 2 Obj: 1   RANDOM: Y

34. Using the following sampling distribution, what is the mean of this distribution?

$\overline{X}$	$P(\overline{X})$
7	1/13
12	2/13
15	3/13
20	4/13
28	1/13
30	2/13

Answer: 18.77 Type: S   Sect: 2 Obj: 1   RANDOM: Y

35. Using the following sampling distribution, what is the mean of this distribution?

$\overline{X}$	$P(\overline{X})$
0	3/17
5	1/17
7	4/17
12	2/17
20	1/17
35	5/17
50	1/17

Answer: 17.76 Type: S   Sect: 2 Obj: 1   RANDOM: Y

36. What is the sum of the S^2 column for the following data?

Sample	S^2
2,2	
2,5	
2,9	
5,2	
5,5	
5,9	
9,2	
9,5	
9,9	

Answer: 74 Type: S   Sect: 2 Obj: 1   RANDOM: Y

37. Suppose we want to find the maximum value for each of our sample points; call this value MAX. Find the MAX of each of the following values and use it to construct a sampling distribution. The values are: (50, 75), (50, 100), (50, 150), (50, 200), (75, 50), (75, 100), (75, 200), (75, 150), (100, 50), (100, 150), (100, 75), and (100, 200).

Answer:

MAX Value	P(MAX Value)
75	2/12 = 1/6
100	4/12 = 1/3
150	3/12 = 1/4
200	3/12 = 1/4
	1

Type: S   Sect: 2 Obj: 1   RANDOM: Y

38. Use the following random samples: (3, 7), (3, 10), (3, 15), (7, 10), (7, 15), (7, 3), (10, 3), (10, 15), (10, 7). What are the possible values of $\overline{X}$?

Answer: 5, 6.5, 8.5, 9, 11, 12.5 Type: S   Sect: 2 Obj: 1   RANDOM: Y

39. Use the following sampling distribution of $\overline{X}$ to answer the question.

$\overline{X}$	$P(\overline{X})$
3	0.5
5	0.2
7	0.3

What is $P(\overline{X} < 6)$?

Answer: 0.7 Type: S   Sect: 2 Obj: 1   RANDOM: Y

40. Use the following sampling distribution of $\overline{X}$ to answer the question.

$\overline{X}$	$P(\overline{X})$
0	0.05
3	0.21
6	0.16
9	0.32
12	0.18
15	0.08

What is $P(\overline{X}$ is even)?

Answer: 0.39 Type: S   Sect: 2 Obj: 1   RANDOM: Y

41. Use the following sampling distribution of $\overline{X}$ to answer the question.

$\overline{X}$	$P(\overline{X})$
0	0.05
3	0.21
6	0.16
9	0.32
12	0.18
15	0.08

What is $P(3 \leq \overline{X} < 11)$?

Answer: 0.69 Type: S   Sect: 2 Obj: 1   RANDOM: Y

42. Use the following sampling distribution of $\overline{X}$ to answer the question.

$\overline{X}$	$P(\overline{X})$
0	0.05
3	0.21
6	0.16
9	0.32
12	0.18
15	0.08

What is the mean of this sampling distribution?

Answer: 7.83 Type: S   Sect: 2 Obj: 1   RANDOM: Y

Determine if the finite population correction factor is necessary. If it is, calculate it.

43. $N = 600$, $n = 20$

Answer: Not necessary Type: S   Sect: 7 Obj: 3   RANDOM: Y

44. $N = 1,500$, $n = 100$

Answer: Necessary, correction factor = 0.9664 Type: S   Sect: 7 Obj: 3   RANDOM: Y

45. $N = 4,000$, $n = 50$

Answer: Not necessary Type: S   Sect: 7 Obj: 3   RANDOM: Y

46. $N = 250$, $n = 35$

Answer: Necessary, correction factor = 0.9292 Type: S   Sect: 7 Obj: 3   RANDOM: Y

47. $N = 300$, $n = 30$

Answer: Necessary, correction factor = 0.9503   Type: S   Sect: 7   Obj: 3   RANDOM: Y

For the following problem, assume the sample is taken from a large population and the correction factor can be ignored.

48. The mean weight of 10-year-old girls is 80 pounds, and the standard deviation is 6 pounds. If a sample of 31 girls is selected, what is the probability that the mean of the sample is more than 82 pounds?

Answer: 0.0314   Type: S   Sect: 5   Obj: 2   RANDOM: Y

49. The mean weight of 10-year-old girls is 80 pounds, and the standard deviation is 6 pounds. If a sample of 31 girls is selected, what is the probability that the mean of the sample is less than 79 pounds?

Answer: 0.1762   Type: S   Sect: 5   Obj: 2   RANDOM: Y

50. The mean weight of 10-year-old girls is 80 pounds, and the standard deviation is 6 pounds. If a sample of 31 girls is selected, what is the probability that the mean of the sample is between 78.5 and 80.5 pounds?

Answer: 0.5949   Type: S   Sect: 5   Obj: 2   RANDOM: Y

51. The mean grade-point average of the Lambda Chi fraternity at a large university is 2.7 with a standard deviation of 0.27. In a group of 40 fraternity brothers, what is the probability that their grade-point average is less than 2.6?

Answer: 0.0096   Type: S   Sect: 5   Obj: 2   RANDOM: Y

52. The mean grade-point average of the Lambda Chi fraternity at a large university is 2.7 with a standard deviation of 0.27. In a group of 40 fraternity brothers, what is the probability that their grade-point average is more than 2.65?

Answer: 0.8790   Type: S   Sect: 5   Obj: 2   RANDOM: Y

53. The mean grade-point average of the Lambda Chi fraternity at a large university is 2.7 with a standard deviation of 0.27. In a group of 40 fraternity brothers, what is the probability that their grade-point average is less than 2.73?

Answer: 0.758   Type: S   Sect: 5   Obj: 2   RANDOM: Y

54. The mean grade-point average of the Lambda Chi fraternity at a large university is 2.7 with a standard deviation of 0.27. In a group of 40 fraternity brothers, what grade-point average would we expect the top 10% of the fraternity brothers to have?

    Answer: 2.75 Type: S   Sect: 5 Obj: 2   RANDOM: Y

55. A particular region has an annual mean rainfall of 80 inches and a standard deviation of 8 inches. What is the probability that the mean rainfall of 32 selected years is more than 82 inches?

    Answer: 0.0793 Type: S   Sect: 5 Obj: 2   RANDOM: Y

56. A particular region has an annual mean rainfall of 80 inches and a standard deviation of 8 inches. What is the probability that the mean rainfall of 32 selected years is less than 79 inches?

    Answer: 0.2389 Type: S   Sect: 5 Obj: 2   RANDOM: Y

Answer the following question. Be sure and check to see if the finite correction factor is needed.

57. A study of 500 pastors in Baptist churches found that their sermons lasted an average of 20 minutes with a standard deviation of 3 minutes. If a sample of 50 pastors is selected, what is the probability that their sermons will average more than 21 minutes in length?

    Answer: 0.0066 Type: S   Sect: 7 Obj: 3   RANDOM: Y

58. A study of 500 pastors in Baptist churches found that their sermons lasted an average of 20 minutes with a standard deviation of 3 minutes. If a sample of 50 pastors is selected, what is the probability that their sermons will average less than 19.5 minutes in length?

    Answer: 0.1075 Type: S   Sect: 7 Obj: 3   RANDOM: Y

59. A study of 500 pastors in Baptist churches found that their sermons lasted an average of 20 minutes with a standard deviation of 3 minutes. If a sample of 50 pastors is selected, what is the probability that their sermons will average between 19 and 20.5 minutes in length?

    Answer: 0.8859 Type: S   Sect: 7 Obj: 3   RANDOM: Y

60. A survey of 250 lobster fishermen found that they catch an average of 32 pounds of lobster per day with a standard deviation of 4 pounds. If a random sample of 30 lobster fishermen is selected, what is the probability that their average catch is more than 33 pounds?

Answer: 0.0721 Type: S   Sect: 7 Obj: 3   RANDOM: Y

61. Using the following sampling distribution, what is the mean of this distribution?

$\overline{X}$	$P(\overline{X})$
2	1/8
5	1/8
8	3/8
10	1/8
16	1/4

Answer: 9.125 Type: S   Sect: 2 Obj: 1   RANDOM: Y

62. Using the following sampling distribution, answer the question.

$\overline{X}$	$P(\overline{X})$
2	1/8
5	1/8
8	3/8
10	1/8
16	1/4

What is $P(0 \le \overline{X} < 10)$?

Answer: 5/8 Type: S   Sect: 2 Obj: 1   RANDOM: Y

63. If the finite population correction factor is 0.89 and $N = 5,000$, what is the value of n?

Answer: 1,040 Type: S   Sect: 7 Obj: 3   RANDOM: Y

64. Using the following sampling distribution, answer the question.

$\overline{X}$	$P(\overline{X})$
2	1/8
5	1/8
8	3/8
10	1/8
16	1/4

What is $P(\overline{X} \ge 5)$?

Answer: 7/8 Type: S   Sect: 2 Obj: 1   RANDOM: Y

65. Using the following sampling distribution, answer the question.

$\overline{X}$	$P(\overline{X})$
2	1/8
5	1/8
8	3/8
10	1/8
16	1/4

What is $P(\overline{X} < 10)$?

Answer: 5/8   Type: S   Sect: 2   Obj: 1   RANDOM: Y

66. If the finite population correction factor is 0.9 and $N = 10,000$, what is the value of n?

Answer: 592   Type: S   Sect: 7   Obj: 3   RANDOM: Y

67. Using the following sampling distribution, answer the question.

$\overline{X}$	$P(\overline{X})$
1	0.2
5	0.1
8	0.1
16	0.3
25	0.2
40	0.1

What is $P(\overline{X}$ is even)?

Answer: 0.5   Type: S   Sect: 2   Obj: 1   RANDOM: Y

68. Using the following sampling distribution, answer the question.

$\overline{X}$	$P(\overline{X})$
1	0.2
5	0.1
8	0.1
16	0.3
25	0.2
40	0.1

What is $P(\overline{X}$ is a perfect square)?

Answer: 0.7   Type: S   Sect: 2   Obj: 1   RANDOM: Y

69. If the finite population correction factor is 0.78 and $N = 800$, what is the value of n?

Answer: 314   Type: S   Sect: 7   Obj: 3   RANDOM: Y

70. The average 600-sq.-ft. apartment in Japan rents for $675 a month with a standard deviation of $50. If a random sample of 49 apartments is selected, what is the probability that the mean rent is more than $690?

Answer: 0.0179 Type: S   Sect: 5 Obj: 2   RANDOM: Y

71. The average 600-sq.-ft. apartment in Japan rents for $675 a month with a standard deviation of $50. If a random sample of 49 apartments is selected, what is the probability that the mean rent is less than $670?

Answer: 0.2877 Type: S   Sect: 5 Obj: 2   RANDOM: Y

72. The average 600-sq.-ft. apartment in Japan rents for $675 a month with a standard deviation of $50. If a random sample of 49 apartments is selected, what is the probability that the mean rent is between $668 and $680?

Answer: 0.5945 Type: S   Sect: 5 Obj: 2   RANDOM: Y

73. The average 600-sq.-ft. apartment in Japan rents for $675 a month with a standard deviation of $50. If a random sample of 49 apartments is selected, what is the probability that the mean rent is between $670 and $673?

Answer: 0.1477 Type: S   Sect: 5 Obj: 2   RANDOM: Y

74. The average 600-sq.-ft. apartment in Japan rents for $675 a month with a standard deviation of $50. In a sample of 100 apartments, the lowest 7% of the apartment would rent for what amount?

Answer: $667.60 or less Type: S   Sect: 5 Obj: 2   RANDOM: Y

75. A standard test has a mean of 500 and a standard deviation of 40. If this test is given to 1,600 students, what is the probability that the mean of this sample is less than 502?

Answer: 0.9772 Type: S   Sect: 5 Obj: 2   RANDOM: Y

76. A standard test has a mean of 500 and a standard deviation of 60. If this test is given to 1,600 students, what is the probability that the mean of this sample is more than 501?

Answer: 0.2514 Type: S   Sect: 5 Obj: 2   RANDOM: Y

77. A standard test has a mean of 500 and a standard deviation of 60. If this test is given to 1,600 students, what is the probability that the mean of this sample is between 498 and 501.5?

Answer: 0.7495 Type: S   Sect: 5 Obj: 2   RANDOM: Y

78. A standard test has a mean of 500 and a standard deviation of 60. If this test is given to 1,600 students, what is the probability that the mean of this sample is more than 497?

    Answer: 0.9772 Type: S   Sect: 5 Obj: 2   RANDOM: Y

79. A standard test has a mean of 500 and a standard deviation of 60. If this test is given to 2,500 students, what is the probability that the mean of this sample is between 497 and 502?

    Answer: 0.9463 Type: S   Sect: 5 Obj: 2   RANDOM: Y

80. The average number of commercials on TV during a half-hour program was 12 with a standard deviation of 4. If 36 programs are selected at random, what is the probability that the average number of commercials is less than 10?

    Answer: 0.0013 Type: S   Sect: 5 Obj: 2   RANDOM: Y

81. The average number of commercials on TV during a half-hour program was 12 with a standard deviation of 4. If 36 programs are selected at random, what is the probability that the average number of commercials is more than 11.5?

    Answer: 0.7734 Type: S   Sect: 5 Obj: 2   RANDOM: Y

82. The average number of commercials on TV during a half-hour program was 12 with a standard deviation of 4. If 64 programs are selected at random, what is the probability that the average number of commercials is between 11 and 12.5?

    Answer: 0.8185 Type: S   Sect: 5 Obj: 2   RANDOM: Y

83. The average number of commercials on TV during a half-hour program was 12 with a standard deviation of 4. If 64 programs are selected at random, what is the probability that the average number of commercials is less than 13.25?

    Answer: 0.9938 Type: S   Sect: 5 Obj: 2   RANDOM: Y

84. The average number of commercials on TV during a half-hour program was 12 with a standard deviation of 4. Sixty-four programs are selected at random, and you want to watch a program with fewer commercials. How many commercials would there be in the lowest 2% of the programs?

    Answer: 11 or less Type: S   Sect: 5 Obj: 2   RANDOM: Y

85. What percent of the $\overline{X}$ values are within 2 standard errors of the population mean?

    Answer: 95.44% Type: S   Sect: 5 Obj: 2   RANDOM: Y

86. The middle 60% of the $\overline{X}$ values are within how many standard errors of the population mean?

    Answer: 0.84 standard errors Type: S   Sect: 5 Obj: 2   RANDOM: Y

87. What percent of $\overline{X}$ values from large samples are within 1.75 standard errors of the population mean?

    Answer: 91.98% Type: S   Sect: 5 Obj: 2   RANDOM: Y

88. The middle 80% of the $\overline{X}$ values are within how many standard errors of the population mean?

    Answer: 1.28 standard errors Type: S   Sect: 5 Obj: 2   RANDOM: Y

89. The middle 76% of the $\overline{X}$ values are within how many standard errors of the population mean?

    Answer: 1.175 standard errors Type: S   Sect: 5 Obj: 2   RANDOM: Y

90. What percent of $\overline{X}$ values from large samples are within 2.17 standard errors of the population mean?

    Answer: 97% Type: S   Sect: 5 Obj: 2   RANDOM: Y

91. The middle 84% of the X values are within how many standard errors of the population mean?

    Answer: 1.41 standard errors Type: S   Sect: 5 Obj: 2   RANDOM: Y

MULTIPLE CHOICE

```
*******************************************************************
The remaining questions in this chapter require the use of the formula
generation capabilities of ESATEST III.  Please see the ESATEST III
documentation if you need directions on how to use them.
*******************************************************************
```

92. The standard deviation of a population is @15, N = @100, and n = `@25`. Using the finite population correction factor, what is the adjusted standard error of the mean?

    a. 12.61   b. 12.27   c. 11.14   d. 10.52

    Answer: @1/@3^0.5*((@2-@3)/(@2-1))^0.5,15,10,20,0,100,100,120,0,25,20,30,0,[2]   Type: M

    RANDOM: F

93. If N = @1000 and n = @150, what is the finite population correction factor?

    a. | 0.85   b. | 0.92   c. | 0.72   d. | 0.87

Answer: ((@1-@2)/(@1-1))^0.5,1000,1000,1200,0,150,100,200,0,[2]   Type: M   RANDOM: F

94. If the finite correction factor is @0.92 and N = @2500, what is the value of n?

    a. | 103   b. | 385   c. | 1375   d. | 535

Answer: @2-@1^2*(@2-1),0.92,0.8,0.95,2,2500,2000,3000,0,[0]   Type: M   RANDOM: F

SHORT ANSWER

95. Using the following sampling distribution, what is the mean of this distribution?

$\overline{X}$	$P(\overline{X})$
3	@1/11
5	2/11
7	@5/11
10	1/11
15	`8-~1-~2`/11

Answer: (140-12*@1-8*@2)/11,1,1,2,0,5,3,5,0,[2]   Type: S   RANDOM: F

96. Use the following sampling distribution of $\overline{X}$ to answer the question.

$\overline{X}$	$P(\overline{X})$
3	@0.5
5	@0.2
7	`1-~1-~2`

What is $P(\overline{X} < 6)$?

Answer: @1+@2,0.5,0.3,0.6,1,0.2,0.1,0.3,1 Type: S   RANDOM: F

97. Use the following sampling distribution of $\overline{X}$ to answer the question.

$\overline{X}$	$P(\overline{X})$
0	0.05
3	@0.21
6	@0.16
9	0.32
12	`0.55-~1-~2`
15	0.08

What is $P(\overline{X}$ is even)?

Answer: 0.6-@1,0.21,0.2,0.3,2,0.16,0.1,0.2,2 Type: S   RANDOM: F

98. Use the following sampling distribution of $\overline{X}$ to answer the question.

$\overline{X}$	$P(\overline{X})$
0	0.05
3	@0.21
6	@0.16
9	0.32
12	`0.55-~1-~2`
15	0.08

What is $P(3 \leq \overline{X} < 11)$?

Answer: 0.32+@1+@2,0.21,0.2,0.3,2,0.16,0.1,0.2,2 Type: S   RANDOM: F

99. Use the following sampling distribution of $\overline{X}$ to answer the question.

$\overline{X}$	$P(\overline{X})$
0	0.05
3	@0.21
6	@0.16
9	0.32
12	`0.55-~1-~2`
15	0.08

What is the mean of this sampling distribution?

Answer: 10.68-9*@1-6*@2,0.21,0.2,0.3,2,0.16,0.1,0.2,2 Type: S   RANDOM: F

100.Using the following sampling distribution, answer the question.

$\overline{X}$	$P(\overline{X})$
2	@1/8
5	1/8
8	`4-~1`/8
10	1/8
16	1/4

What is $P(0 \leq \overline{X} < 10)$?

Answer: 5/8,1,1,3,0 Type: S   RANDOM: F

101.Using the following sampling distribution, answer the question.

$\overline{X}$	$P(\overline{X})$
2	@1/8
5	1/8
8	`4-~1`/8
10	1/8
16	1/4

What is $P(\overline{X} \geq 5)$?

Answer: `(8-~1)/8`,1,1,3,0 Type: S   RANDOM: F

102.Using the following sampling distribution, answer the question.

$\overline{X}$	$P(\overline{X})$
1	0.2
5	@0.1
8	0.1
16	@0.3
25	`0.6-~1-~2`
40	0.1

What is the $P(\overline{X}$ is a perfect square)?

Answer: 0.8-@1,0.1,0.1,0.2,1,0.3,0.1,0.3,1 Type: S   RANDOM: F

MULTIPLE CHOICE

103. The mean of a population is @400 and the standard deviation is `@50`. A sample of size 225 is selected. What is the standard error of the mean?

   a. 10.22   b. 13.33   c. 11.78   d. 18

   Answer: @2/15,400,300,500,0,50,40,60,0,[2]   Type: M   RANDOM: F

SHORT ANSWER

104. What is the sum of the S^2 column for the following data?

<u>Sample</u>	<u>S^2</u>
@2,~1	
~1,@5	
~1,@9	
~2,~1	
~2,~2	
~2,~3	
~3,~1	
~3,~2	
~3,~3	

   Answer: (@2-@1)^2+(@3-@1)^2+(@2-@3)^2,2,1,4,0,5,5,8,0,9,9,15,0 Type: S   RANDOM: F

105. Use the following random samples: (@3, 7), (~1, 10), (~1, 15), (7, 10), (7, 15), (7,~1), (10,~1), (10, 15), (10, 7). What are the possible values of $\overline{X}$?

   Answer: `(~1+7)/2` , `(~1+10)/2` , 8.5, `(~1+15)/2` , 11, 12.5,3,3,6,0 Type: S

   RANDOM: F

106. Determine if the finite population correction factor is necessary. If it is, calculate it. $N = $ @600, $n = $ @20

   Answer: Not necessary,600,600,700,0,20,10,20,0 Type: S   RANDOM: F

CHAPTER 8

TRUE/FALSE

1. The best point estimate of the population standard deviation is the sample mean.

   Answer: F Type: T Sect: 3 Obj: 1   RANDOM: Y

2. A 95% confidence interval is larger than a 90% confidence interval.

   Answer: T Type: T Sect: 2 Obj: 1   RANDOM: Y

3. Cheryl wants to calculate the 99% confidence interval around the mean using the t-distribution when $n = 15$. The critical value she would use is 2.977.

   Answer: T Type: T Sect: 3 Obj: 2   RANDOM: Y

4. Fred wants to calculate the 98% confidence interval around the mean using the z-distribution. His sample of 64 items has a mean of 35 and a standard deviation of 6. The maximum error of estimate would be 0.218.

   Answer: F Type: T Sect: 7 Obj: 3   RANDOM: Y

5. One property of a good estimator is that it is relatively efficient.

   Answer: T Type: T Sect: 1 Obj: 1   RANDOM: Y

6. The dean wants to estimate the average number of hours taken by students at Bison University. The standard deviation is thought to be 2.5 hours. If the dean wants to be 95% confident of finding the true mean within 1.5 hours, he must sample at least 11 students.

   Answer: T Type: T Sect: 7 Obj: 4   RANDOM: Y

7. Delores wants to construct a 94% confidence interval around the mean using the z-score. Her critical value would be 1.88.

   Answer: T Type: T Sect: 2 Obj: 2   RANDOM: Y

8. A sample of 400 applicants for law school included 75 women. The 98% confidence interval of the true proportion of women who applied would be from 0.16 to 0.22.

   Answer: F Type: T Sect: 10 Obj: 5   RANDOM: Y

9. A confidence interval was constructed around a proportion. The interval was from 23.6% to 34.2%. The proportion used to construct this interval was 28.9%.

Answer: T Type: T   Sect: 10 Obj: 5   RANDOM: Y

10. A retailer wants to estimate with 99% confidence the number of people who buy at his store. A previous study showed that 15% of those interviewed had shopped at his store. He wishes to be accurate within 3% of the true proportion. The minimum sample size necessary would be 943.

Answer: T Type: T   Sect: 11 Obj: 6   RANDOM: Y

MULTIPLE CHOICE

11. What is the best point estimate of the mean of the population?
    a. mode of the sample
    b. mean of the sample
    c. median of the sample
    d. standard deviation of the sample

Answer: b Type: M   Sect: 1 Obj: 1   RANDOM: Y

12. Which of the following percents would be the smallest confidence interval?
    a. 99%
    b. 95%
    c. 90%
    d. 98%

Answer: c Type: M   Sect: 2 Obj: 1   RANDOM: Y

13. John wants to find a 95% confidence interval around the mean using the t-distribution when $n = 18$. What would the critical value be?
    a. 2.101
    b. 2.093
    c. 2.110
    d. 1.740

Answer: c Type: M   Sect: 3 Obj: 2   RANDOM: Y

14. A sample of the ACT scores of 50 students in Basic Math revealed a mean score of 10 with a standard deviation of 3. What is the 98% confidence interval of the mean for these students?
    a. 9.13-10.87
    b. 9.86-10.14
    c. 9.01-10.99
    d. 9.75-10.75

Answer: c Type: M   Sect: 3 Obj: 3   RANDOM: Y

15. A weatherman kept records over 16 days of the amount of rainfall in his area. The average rainfall was 1.6 inches with a standard deviation of 0.35 inch. What is the 90% confidence interval of the mean?
    a. 1.447-1.753
    b. 1.456-1.744
    c. 1.414-1.786
    d. 1.562-1.638

    Answer: a Type: M   Sect: 3 Obj: 3   RANDOM: Y

16. A drag racer wants to estimate the average number of laps he can go on one set of tires. He wishes to be accurate to within 1.5 laps. From a previous study he knows that the standard deviation is 5 laps. How large a sample must he select if he wants to be 98% confident of finding the true mean?
    a. 60
    b. 61
    c. 47
    d. 74

    Answer: b Type: M   Sect: 7 Obj: 4   RANDOM: Y

17. A sample of 76 people was used to estimate with 99% confidence the mean. The standard deviation was 6. What was the maximum error of estimate?
    a. 0.56
    b. 0.75
    c. 1.77
    d. 3.15

    Answer: c Type: M   Sect: 7 Obj: 4   RANDOM: Y

18. It was found that in a sample of 80 teenage boys, 70% had received speeding tickets. What is the 95% confidence interval of the true proportion of teenage boys who have received speeding tickets?
    a. 0.60-0.80
    b. 0.58-0.82
    c. 0.62-0.78
    d. 0.59-0.81

    Answer: a Type: M   Sect: 10 Obj: 5   RANDOM: Y

19. A political candidate wants to estimate his chances of winning the coming election for mayor to within 2% with 99% confidence. He believes that 54% of the people will vote for him. What is the minimum sample size necessary?
    a. 4,138
    b. 3,372
    c. 4,134
    d. 4,160

    Answer: c Type: M   Sect: 11 Obj: 6   RANDOM: Y

20. A sample of 2,172 was used to estimate a proportion with 98% confidence.   If $p = 0.5$, what was the amount of error?
    a. 0.025
    b. 0.035
    c. 0.028
    d. 0.032

    Answer: a Type: M   Sect: 11 Obj: 6   RANDOM: Y

SHORT ANSWER

21. What is the critical value for a 96% confidence interval, using the standard normal distribution?

    Answer: 2.05 Type: S   Sect: 3 Obj: 2   RANDOM: Y

22. What is the critical value for a 98% confidence interval, using the standard normal distribution?

    Answer: 2.33 Type: S   Sect: 3 Obj: 2   RANDOM: Y

23. What is the critical value for a 92% confidence interval, using the standard normal distribution?

    Answer: 1.75 Type: S   Sect: 3 Obj: 2   RANDOM: Y

24. What is the critical value for a 98% confidence interval, using a t-distribution with $n = 18$?

    Answer: 2.567 Type: S   Sect: 3 Obj: 2   RANDOM: Y

25. What is the critical value for a 99% confidence interval, using a t-distribution with $n = 25$?

    Answer: 2.797 Type: S   Sect: 3 Obj: 2   RANDOM: Y

26. What is the critical value for a 95% confidence interval, using a t-distribution with $n = 15$?

    Answer: 2.145 Type: S   Sect: 3 Obj: 2   RANDOM: Y

27. What is the critical value for a 90% confidence interval, using a t-distribution with $n = 28$?

    Answer: 1.703 Type: S   Sect: 3 Obj: 2   RANDOM: Y

28. What is the critical value for a 95% confidence interval, using a
    t-distribution with $n = 20$?

    Answer: 2.093 Type: S   Sect: 3 Obj: 2   RANDOM: Y

29. What is the critical value for a 98% confidence interval, using a
    t-distribution with 9 degrees of freedom?

    Answer: 2.821 Type: S   Sect: 3 Obj: 2   RANDOM: Y

30. A sample of 60 faculty members at a large university found that their average taxes were
    $500 a month with a standard deviation of $35. What is the 90% confidence interval of
    the average taxes of all faculty members at the university?

    Answer: $492.57 \leq \mu \leq $507.43 Type: S   Sect: 3 Obj: 3   RANDOM: Y

31. In a recent survey of 75 homemakers, the average number of hours per week they watched
    television was 28 with a standard deviation of 3 hours. What is the 99% confidence
    interval of the mean number of hours watched by all homemakers?

    Answer: $27.11 \leq \mu \leq 28.89$ Type: S   Sect: 3 Obj: 3   RANDOM: Y

32. At a diet center, a sample of 25 people consumed an average of 1,500 calories a day with
    a standard deviation of 100 calories. What is the 98% confidence interval of the mean
    number of calories for all dieters at this center?

    Answer: $1,450.16 \leq \mu \leq 1,549.84$ Type: S   Sect: 3 Obj: 3   RANDOM: Y

33. In the city of Topeka, a researcher found that the average utility bill of 120 residents
    was $75 with a standard deviation of $10. What is the 95% confidence interval of the
    mean for all Topeka residents?

    Answer: $73.21 \leq \mu \leq 76.79$ Type: S   Sect: 3 Obj: 3   RANDOM: Y

34. The owner of a pet shop wanted to estimate the average number of fish her customers
    purchased. In a sample of 20 customers, she found that they purchased an average of 10
    fish with a standard deviation of 2 fish. What is the 95% confidence interval of the
    true mean?

    Answer: $9.064 \leq \mu \leq 10.936$ Type: S   Sect: 3 Obj: 3   RANDOM: Y

35. The average number of miles that 80 truckers drove in a day was 550 with a standard
    deviation of 50 miles. What is the 98% confidence interval of the true mean number of
    miles driven by all truckers?

    Answer: $536.975 \leq \mu \leq 563.025$ Type: S   Sect: 3 Obj: 3   RANDOM: Y

36. The average number of textbooks received per year by 100 college professors at Washington University is 30 with a standard deviation of 3. What is the 99% confidence interval of the true mean?

Answer: $29.23 \leq \mu \leq 30.77$ Type: S   Sect: 3 Obj: 3   RANDOM: Y

37. At a large factory, the foreman wanted to estimate the average number of chicken dinners that could be produced in an hour. A sample of 28 employees revealed an average production rate of 70 dinners per hour with a standard deviation of 4 dinners. What is the 98% confidence interval of the true mean?

Answer: $68.131 \leq \mu \leq 71.869$ Type: S   Sect: 3 Obj: 3   RANDOM: Y

38. Thirty-six new students enrolled at a private school had an average entrance exam score of 85 with a standard deviation of 8. What is the 96% confidence interval of the true mean of all students enrolling at this school?

Answer: $82.27 \leq \mu \leq 87.73$ Type: S   Sect: 3 Obj: 3   RANDOM: Y

39. A random sample of 45 cups of coffee from a coffee dispenser revealed a mean of 7.95 ounces with a standard deviation of 0.15 ounce. What is the 99% confidence interval of the true mean of all cups coming from this dispenser?

Answer: $7.892 \leq \mu \leq 8.008$ Type: S   Sect: 3 Obj: 3   RANDOM: Y

40. A random sample of 16 women in a beginning aerobics class had a mean weight of 142 pounds and a standard deviation of 12 pounds. What is the 95% confidence interval of the mean of all women in the aerobics class?

Answer: $135.607 \leq \mu \leq 148.393$ Type: S   Sect: 3 Obj: 3   RANDOM: Y

41. A district attorney in Los Angeles wants to determine with 98% confidence the average number of days it takes for a murder suspect to come up for trial. If he wants to be accurate to within 4 days, how large a sample does he need? (It is thought that the standard deviation is 7 days.)

Answer: 17 Type: S   Sect: 7 Obj: 4   RANDOM: Y

42. A nutritionist wants to estimate the mean amount of cholesterol in a certain variety of chicken eggs. She wants to be within 14 mg with 99% confidence. It is believed that the standard deviation is 25 mg. How large a sample of eggs should she use?

Answer: 22 Type: S   Sect: 7 Obj: 4   RANDOM: Y

43. A researcher wants to estimate the average number of hours men spend watching television on the weekend. He wants to be 98% sure his estimate is correct. If the standard deviation is 2 hours, how large a sample is necessary to be accurate to within 1/2 hour?

Answer: 87 Type: S   Sect: 7 Obj: 4   RANDOM: Y

44. In order to determine sales figures, a car dealer wants to find the 95% confidence interval of the mean sales price of his models. He wants to be accurate to within $150. If the standard deviation of the cars is thought to be $600, how large a sample must he select?

Answer: 62 Type: S   Sect: 7 Obj: 4   RANDOM: Y

45. A tornado has blown through a section of Collinsville. The insurance adjuster wants to estimate the average claim to within $300. If the standard deviation is believed to be $1,000 and the company wants to be 99% sure of its figures, how large a sample must be taken?

Answer: 74 Type: S   Sect: 7 Obj: 4   RANDOM: Y

46. A nurseryman wants to estimate how many poinsettias he will need for Christmas. He estimates the standard deviation to be 15. If he wishes to be 95% certain of having enough plants, how large a sample must he take to be within 5 plants?

Answer: 35 Type: S   Sect: 7 Obj: 4   RANDOM: Y

47. A clothing store owner wants to determine the 98% confidence of the average price of men's dress shirts. She wants to be accurate to within $1.50. If the standard deviation is $4.00, how large a sample must be selected to obtain the desired information?

Answer: 39 Type: S   Sect: 7 Obj: 4   RANDOM: Y

48. For 12 years, a meteorologist kept track of the number of days it snowed in February. The results were as follows: 4, 5, 12, 2, 9, 10, 6, 5, 8, 15, 2, 9. What is the 98% confidence interval of the true mean?

Answer: $4.721 \leq \mu \leq 9.779$ Type: S   Sect: 3 Obj: 3   RANDOM: Y

49. Ten dieters kept track of how many glasses of water they drank each day. Their data were 4, 6, 7, 8, 5, 7, 4, 8, 3, 5. What is the 95% confidence interval of the true mean?

Answer: $4.436 \leq \mu \leq 6.964$ Type: S   Sect: 3 Obj: 3   RANDOM: Y

50. The weights (in pounds) of 15 models were recorded as follows: 105, 110, 120, 118, 105, 108, 110, 117, 125, 105. 113. 105, 119, 121, and 120. What is the 99% confidence interval of the true mean of all models?

    Answer: $108.045 \leq \mu \leq 118.355$ Type: S   Sect: 3 Obj: 3   RANDOM: Y

51. A class of 15 second-grade students was given a test to see how many simple addition problems they could do in 3 minutes. The results were as follows: 15, 25, 32, 18, 23, 35, 16, 23, 30, 19, 27, 33, 17, 29, 23. What is the 90% confidence interval of the true mean?

    Answer: $21.37 \leq \mu \leq 27.29$ Type: S   Sect: 3 Obj: 3   RANDOM: Y

52. The average lifetime of 64 19-inch televisions was 6.2 years with a standard deviation of 0.8 years. What is the 98% confidence interval of the true mean of all similar television sets?

    Answer: $5.97 \leq \mu \leq 6.43$ Type: S   Sect: 3 Obj: 3   RANDOM: Y

53. The average lifetime of 18 televisions was 6.3 years with a standard deviation of 0.7 years. What is the 99% confidence interval of the true mean of all similar television sets?

    Answer: $5.82 \leq \mu \leq 6.78$ Type: S   Sect: 3 Obj: 3   RANDOM: Y

54. A researcher wants to estimate with 98% confidence the number of people who will have back pain this year. He does not know the standard deviation but he wants to be within 1/3 standard deviation of the true mean. How large a sample should be taken?

    Answer: 49 Type: S   Sect: 7 Obj: 4   RANDOM: Y

55. A sample of 300 racing cars showed that 90 cost over $500,000. What is the 99% confidence interval of the true proportion of cars costing over $500,000?

    Answer: $0.232 \leq p \leq 0.368$ Type: S   Sect: 10 Obj: 5   RANDOM: Y

56. Pizza Inn wanted to determine what proportion of its customers ordered only cheese pizza. Out of 90 customers, 11 ordered cheese pizza. What is the 98% confidence interval of the true proportion of customers who order only cheese pizza?

    Answer: $0.042 \leq p \leq 0.202$ Type: S   Sect: 10 Obj: 5   RANDOM: Y

57. In a sample of 400 Americans over age 21, 75% said they drink alcoholic beverages. What is the 95% confidence interval of the true proportion of all Americans who drink alcoholic beverages?

    Answer: $0.708 \leq p \leq 0.792$ Type: S   Sect: 10 Obj: 5   RANDOM: Y

58. The manager of a dairy store wants to determine the proportion of people who buy milk each day. During one day, 150 out of 250 people bought milk. What is the 95% confidence interval for the true proportion of customers who buy milk each day?

    Answer: $0.54 \leq p \leq 0.66$ Type: S   Sect: 10 Obj: 5   RANDOM: Y

59. In a large company, 200 managers out of a sample of 250 work more than 40 hours a week. What is the 96% confidence interval for the true proportion of all managers who work over 40 hours a week?

    Answer: $0.748 \leq p \leq 0.852$ Type: S   Sect: 10 Obj: 5   RANDOM: Y

60. A recent survey at Sun Valley found that 80 out of 120 vacationers chose the resort because of its good ski slopes. What is the 99% confidence interval for the proportion of all vacationers who choose Sun Valley for its ski slopes?

    Answer: $0.556 \leq p \leq 0.778$ Type: S   Sect: 10 Obj: 5   RANDOM: Y

61. A checker at a grocery store found that out of 1,250 sales, 210 were made with food stamps. What is the 96% confidence interval for the true proportion of all food that is bought with food stamps?

    Answer: $0.146 \leq p \leq 0.19$ Type: S   Sect: 10 Obj: 5   RANDOM: Y

62. A random sample of 800 moviegoers found that 760 had seen "Gone with the Wind." What is the 98% confidence interval for the true proportion of moviegoers that have seen this film?

    Answer: $0.932 \leq p \leq 0.968$ Type: S   Sect: 10 Obj: 5   RANDOM: Y

63. A random sample of 200 five-year-old children found that 185 believed in Santa Claus. What is the 95% confidence interval for the true proportion of five-year-olds who believe in Santa Claus?

    Answer: $0.888 \leq p \leq 0.962$ Type: S   Sect: 10 Obj: 5   RANDOM: Y

64. A researcher wants to determine with 95% confidence the number of people who own a home computer. A previous study showed that 27% of those interviewed owned a home computer. What is the minimum sample size necessary to estimate the true proportion of people who own home computers? The researcher wants to be accurate to within 3%.

    Answer: 842 Type: S   Sect: 11 Obj: 6   RANDOM: Y

65. It was reported that 76% of all the land in Botswana was used for agriculture. How many acres would an economist have to test in order to estimate with 95% confidence the true proportion of land that is used for agriculture? He wants to be accurate to within 2%.

    Answer: 1,752 Type: S   Sect: 11 Obj: 6   RANDOM: Y

66. A survey shows that 26% of the citizens of New Zealand belong to the Church of England. A rector wants to know how large a sample he would need in order to estimate with 99% confidence the true proportion of citizens belonging to the Church of England. He wants to be accurate within 2 1/2%. What is the minimum sample size necessary?

    Answer: 2,042 Type: S   Sect: 11 Obj: 6   RANDOM: Y

67. The manager of a supermarket wants to estimate the proportion of customers who will be using food stamps at his store. How large a sample is required to estimate the true proportion to within 2% with 99% confidence?

    Answer: 4,145 Type: S   Sect: 11 Obj: 6   RANDOM: Y

68. A college administrator wants to estimate the proportion of students who attend basketball games at his college. A sample of 300 students revealed that 45 attend the games. How large a sample is needed to estimate with 98% confidence the true proportion of students who attend basketball games? He wants to be accurate to within 3%.

    Answer: 770 Type: S   Sect: 11 Obj: 6   RANDOM: Y

69. A quality control expert wants to estimate the proportion of defective components that are being manufactured by his company to within 3.5%. A sample of 400 components showed that 15 were defective. How large a sample is needed to estimate the true proportion of defective components with 99% confidence?

    Answer: 196 Type: S   Sect: 11 Obj: 6   RANDOM: Y

70. A survey of 120 new truck buyers showed that 30% preferred a red pickup. What is the 95% confidence interval of the true proportion of new truck buyers who prefer a red truck?

    Answer: $0.218 \leq p \leq 0.382$ Type: S   Sect: 10 Obj: 5   RANDOM: Y

71. In Bay Ridge, a survey of 400 people found that 28% had cable TV. What is the 90% confidence interval of the true proportion of people in Bay Ridge who have cable television?

    Answer: $0.243 \leq p \leq 0.317$ Type: S   Sect: 10 Obj: 5   RANDOM: Y

72.A medical researcher wishes to determine the percentage of people who were helped by a new drug. A sample of 80 patients who had taken the new drug revealed that 33 had improved as a result of taking the drug. What is the 95% confidence interval for the true proportion of people who would be helped by using this new drug?

Answer: $0.305 \leq p \leq 0.521$ Type: S   Sect: 10 Obj: 5   RANDOM: Y

73.The college admissions office wishes to estimate the percentage of students over 30 years old who are attending their school. How large should the sample size be in order to be 90% confident that the estimate is within 4% of the true proportion? It is believed that 30% of the student body is over 30 years old.

Answer: 356 Type: S   Sect: 11 Obj: 6   RANDOM: Y

74.A researcher wishes to be 99% confident that her estimate of the true proportion of illiterate persons in Hong Kong is within 3%. What sample size is necessary if a sample of 200 persons showed that 38 were illiterate?

Answer: 1,134 Type: S   Sect: 11 Obj: 6   RANDOM: Y

75.How large a sample must be taken to estimate within 3 1/2% the true proportion of the number of college students who study at least 2 hours a day? We want to be 95% confident of our result.

Answer: 784 Type: S   Sect: 11 Obj: 6   RANDOM: Y

76.How many students should be in a random sample to be 92% confident the error is at most 0.05?

Answer: 307 Type: S   Sect: 11 Obj: 6   RANDOM: Y

77.A newspaper reports that 8% of people it surveyed had 2 jobs. How large a sample is required to be 95% confident of this result within 2.5%?

Answer: 453 Type: S   Sect: 11 Obj: 6   RANDOM: Y

78.A magazine article reports that 25% of its readers are overweight. How large a sample is required to be 98% confident of this statement to within 2%?

Answer: 2,545 Type: S   Sect: 11 Obj: 6   RANDOM: Y

79.A survey reports that 37% of all shareholders in the United States are women. How large a sample is required to be 94% confident of this survey to within 3%?

Answer: 916 Type: S   Sect: 11 Obj: 6   RANDOM: Y

MULTIPLE CHOICE

```
***************************************************************************
```
The remaining questions in this chapter require the use of the formula
generation capabilities of ESATEST III.   Please see the ESATEST III
documentation if you need directions on how to use them.
```
***************************************************************************
```

80. A sample of @2172 was used to estimate a proportion with 98% confidence.   If p = @0.5, what was the amount of error?
    a. |0.025
    b. |0.035
    c. |0.028
    d. |0.032

Answer: 2.33*(@2*(1-@2)/@1)^0.5,2172,2000,3000,0,0.5,0.3,0.7,1,[3]   Type: M

RANDOM: F

CHAPTER 9

TRUE/FALSE

1. A researcher thinks a person's pulse rate will be decreased by exercise. Her alternative hypothesis would contain a greater-than sign.

   Answer: F Type: T   Sect: 5 Obj: 1   RANDOM: Y

2. If the null hypothesis is not rejected when it is false, a Type II error has been committed.

   Answer: T Type: T   Sect: 2 Obj: 2   RANDOM: Y

3. The alternative hypothesis $\mu < 40$ is to be tested at $\alpha = 0.025$ using the z-test. The critical value would be 1.96.

   Answer: F Type: T   Sect: 7 Obj: 3   RANDOM: Y

4. In a two-tailed z-test with $\alpha = 0.02$, the test value was 2.07. The decision would be not to reject the null hypothesis.

   Answer: T Type: T   Sect: 6 Obj: 4   RANDOM: Y

5. A scientist reports that the average number of hours a person can go without sleep and still function normally is 36 with a standard deviation of 4 hours. A sample of 49 people had an average of 38 hours that they could go without sleep and still function normally. The test statistic for this hypothesis would be 3.5.

   Answer: T Type: T   Sect: 6 Obj: 4   RANDOM: Y

6. When the significant level increases, the rejection area decreases.

   Answer: F Type: T   Sect: 2 Obj: 2   RANDOM: Y

7. The critical value for $\alpha = 0.02$ with d.f. = 21 for a two-tailed t-test would be 2.831.

   Answer: F Type: T   Sect: 7 Obj: 6   RANDOM: Y

8. A career counselor claims the average number of years of schooling for an engineer is 17.5. A sample of 16 engineers had a mean of 17 years and a standard deviation of 1.5 years. The test statistic would be -1.33.

   Answer: T Type: T   Sect: 7 Obj: 7   RANDOM: Y

9.A lawyer estimated that 30% of his clients come to him from personal referral. To test his claim, he selected a sample of 40 clients and found that 32% had come to him through personal referrals. The test statistic for this hypothesis would be -0.27.

Answer: F Type: T   Sect: 8 Obj: 8   RANDOM: Y

10.A one-tailed left hypothesis test was performed. The test statistic was -1.98. The p-value for these data would be .0239.

Answer: T Type: T   Sect: 11 Obj: 9   RANDOM: Y

MULTIPLE CHOICE

11.A garbage collector believes that he picks up an average of more than 3 tons of garbage a day. What is the alternative hypothesis for his statement?
 a. $H_1 : \mu > 3$
 b. $H_1 : \mu < 3$
 c. $H_1 : \mu = 3$
 d. $H_1 : \mu \neq 3$

Answer: a Type: M   Sect: 5 Obj: 1   RANDOM: Y

12.A conjecture about a population that may or may not be true is called a
 a. critical value
 b. test value
 c. statistical hypothesis
 d. decision

Answer: c Type: M   Sect: 1 Obj: 2   RANDOM: Y

13.What is another name for a Type II error?
 a. alpha
 b. beta
 c. gamma
 d. correct decision

Answer: b Type: M   Sect: 2 Obj: 2   RANDOM: Y

14.For a one-tailed left z-test with $\alpha = 0.07$, what is the critical value?
 a. -1.81
 b. -1.47
 c. 1.48
 d. -1.48

Answer: d Type: M   Sect: 6 Obj: 3   RANDOM: Y

15. A machine that fills milk bottles is supposed to have a mean amount of milk equal to 32 ounces with a standard deviation of 0.04 ounces. Forty-nine bottles were randomly checked and their mean was found to be 31.98 ounces. What is the test statistic?
    a. 3.5
    b. -3.5
    c. -3.25
    d. 3.25

    Answer: b Type: M   Sect: 6 Obj: 4   RANDOM: Y

16. When you use a smaller value for α, what type of error becomes more likely?
    a. Type I   b. Type II   c. Type III   d. no error

    Answer: b Type: M   Sect: 3 Obj: 2   RANDOM: Y

17. What is the critical value for $\alpha = 0.05$ with d.f. $= 24$ for a one-tailed right t-test?
    a. 2.064
    b. 2.060
    c. 1.708
    d. 1.711

    Answer: d Type: M   Sect: 7 Obj: 6   RANDOM: Y

18. An airline says that it books an average of 75 people on one plane trip. A sample of 9 trips showed a mean of 80 and a standard deviation of 7. What is the test statistic?
    a. 2.575
    b. 2.143
    c. 1.47
    d. 2.347

    Answer: b Type: M   Sect: 7 Obj: 7   RANDOM: Y

19. A teacher claims that more than 30% of her students dropped out last semester. What would be the alternative hypothesis for this claim?
    a. $H_1 : \mu > 0.30$
    b. $H_1 : \mu < 0.30$
    c. $H_1 : p > 0.30$
    d. $H_1 : p < 0.30$

    Answer: c Type: M   Sect: 8 Obj: 9   RANDOM: Y

20. A two-tailed test was conducted to determine if the mean age of gymnasts was equal to 16. The test value was 2.04 and $\alpha = 0.05$. What is the p-value?
    a. 0.0414
    b. 0.0207
    c. 0.4793
    d. 0.0250

    Answer: a Type: M   Sect: 11 Obj: 9   RANDOM: Y

SHORT ANSWER

Use the steps in hypothesis testing to solve the problem.

21. A calculus student claims that the average calculus textbook costs at least $60. The standard deviation is $3.00. A sample of 36 books has an average cost of $62. At $\alpha = 0.01$, test the student's claim.

    Answer:
    $H_0 : \mu \le 60$; $H_1 : \mu > 60$
    C.V. $= 2.33$; $z = 4$
    Reject H_0; the student's claim is correct.

    Type: S   Sect: 6 Obj: 4   RANDOM: Y

22. The manager of a pizza parlor claims his employees make an average of $4.50 per hour with a standard deviation of $0.75. Bob does not believe his claim, so he takes a sample of 40 employees and finds their average salary is $4.25. Test the manager's claim at $\alpha = 0.05$.

    Answer:
    $H_0 : \mu = 4.50$; $H_1 : \mu \ne 4.50$
    C.V. $= \pm 1.96$; $z = -2.11$
    Reject H_0; the manager's claim appears to be incorrect.

    Type: S   Sect: 6 Obj: 4   RANDOM: Y

23. The manager of a pizza parlor claims his employees make an average of $4.50 per hour with a standard deviation of $0.75. Bob does not believe his claim, so he takes a sample of 50 employees and finds their average salary is $4.25. Test the manager's claim at $\alpha = 0.02$.

    Answer:
    $H_0 : \mu = 4.50$; $H_1 : \mu \ne 4.50$
    C.V. $= \pm 2.33$; $z = -2.36$
    Reject H_0; the manager's claim appears to be incorrect.

    Type: S   Sect: 6 Obj: 4   RANDOM: Y

24. A car company claims that the average price of its new car is less than $11,500 with a standard deviation of $700. You feel that the average price is more than the company claims, so you check the price stickers on 32 cars and find the average price is $12,200. At $\alpha = 0.05$, test the car company's claim.

Answer:
$H_0 : \mu \le 11,500$; $H_1 : \mu > 11,500$
C.V. = 1.645; $z = 5.66$
Reject H_0; it appears that the car company's claim is incorrect.

Type: S   Sect: 6   Obj: 4   RANDOM: Y

25. A nutritionist claims that the average number of calories in a serving of popcorn is 70 with a standard deviation of 6. A sample of 50 servings of popcorn yields an average of 72 calories. Check the nutritionist's claim at $\alpha = 0.05$.

Answer:
$H_0 : \mu = 70$; $H_1 : \mu \ne 70$
C.V. = ±1.96; $z = 2.36$
Reject H_0; it appears that the average number of calories of popcorn is not 70.

Type: S   Sect: 6   Obj: 4   RANDOM: Y

26. A manufacturer claims that its television will have an average lifetime of at least 5 years. The standard deviation is 8 months. Sixty-four sets are selected at random, and their average lifetime was found to be 59 months. Is the manufacturer correct? Use $\alpha = 0.025$.

Answer:
$H_0 : \mu \ge 60$; $H_1 : \mu < 60$
C.V. = -1.96; $z = -1$
Do not reject H_0; it appears that the manufacturer's claim is correct and the television will have an average lifetime of at least 5 years.

Type: S   Sect: 6   Obj: 4   RANDOM: Y

27. The average number of colds a child will get per year is 6 with a standard deviation of 1. It was felt that children from a rural community would have fewer colds per year than the average. A group of 64 children were selected, and they had an average of 5 colds per year. Does it appear that rural children have fewer colds? Let $\alpha = 0.01$.

Answer:
$H_0 : \mu \ge 6$; $H_1 : \mu < 6$
C.V. = -2.33; $z = -8$
Reject H_0; it appears that children from rural communities have fewer colds.

Type: S   Sect: 6   Obj: 4   RANDOM: Y

28. A Caribbean island advertises its average summer temperature is 75° with a standard deviation of 5°. Thirty-five days were picked at random and their temperatures were recorded as follows:

```
70  65  82  73  76  81  80  90  66  82  77  76
75  92  85  79  72  75  83  72  77  79  85  87
82  78  89  91  74  78  78  88  90  76  75
```

Let $\alpha = 0.05$ and check the island's claim.

Answer:
$H_0 : \mu = 75$; $H_1 : \mu \neq 75$
C.V. $= \pm 1.96$; $z = 3.76$
Reject H_0; the average summer temperature is different from 75°.

Type: S   Sect: 6 Obj: 4   RANDOM: Y

29. To justify raising car insurance rates, an insurance company claims that the mean accident expense is at least $1,250 per year with a standard deviation of $125. In a survey of 100 randomly selected accident reports, it was found that the mean accident expense was $1,220. At $\alpha = 0.025$, is there evidence that the insurance company is misinformed?

Answer:
$H_0 : \mu \geq 1250$; $H_1 : \mu < 1250$
C.V. $= -1.96$; $z = -2.4$
Reject H_0, it appears the company is misinformed.

Type: S   Sect: 6 Obj: 4   RANDOM: Y

30. A manufacturer of light bulbs claims its light bulbs have a mean life of 1,150 or more hours with a standard deviation of 80 hours. A random testing of 144 light bulbs yielded a mean of 1,125 hours. Is there reason to doubt the manufacturer's claim? Use $\alpha = 0.05$.

Answer:
$H_0 : \mu \geq 1150$; $H_1 : \mu < 1150$
C.V. $= -1.645$; $z = -3.75$
Reject H_0; there is reason to doubt the manufacturer's claim that its light bulbs have a mean life of 1,150 hours or more.

Type: S   Sect: 6 Obj: 4   RANDOM: Y

31. A child psychologist claims that the average number of mistakes an eight-year-old will make on a skills assessment test is 6 with a standard deviation of 2. A random sample of 49 eight-year-olds had a mean of 7 mistakes. Test the psychologist's claim at $\alpha = 0.02$.

Answer:
$H_0 : \mu = 6$; $H_1 : \mu \neq 6$
C.V. $= \pm 2.33$; $z = 3.5$
Reject H_0; the average number of mistakes appears to be different from 6.

Type: S   Sect: 6 Obj: 4   RANDOM: Y

32. A manufacturer claims that its television will have an average lifetime of at least 7 years. The standard deviation is 8 months. Sixty-four sets are selected at random, and their average lifetime was found to be 86 months. Is the manufacturer correct? Use $\alpha = 0.01$.

Answer:
$H_0 : \mu \leq 84$; $H_1 : \mu > 84$
C.V. $= 2.33$; $z = 2$
Do not reject H_0; the average lifetime of the manufacturer's television appears to be less than 7 years.

Type: S   Sect: 6 Obj: 4   RANDOM: Y

33. The chamber of commerce wants to reduce some of the bus routes in town. They feel that the route going to Park Avenue has an average of fewer than 30 passengers on board. To test this claim, several members of the chamber of commerce rode the route to Park Avenue and observed how many passengers were on the bus. Out of 36 trips on the route, the average number of passengers on the Park Avenue route was 26 with a standard deviation of 5. Using a significance level of 0.025, would the chamber of commerce eliminate the Park Avenue route?

Answer:
$H_0 : \mu \geq 30$; $H_1 : \mu < 30$
C.V. $= -1.96$; $z = -4.8$
Reject H_0; the average number of passengers is less than 30. The chamber of commerce might want to eliminate the Park Avenue route.

Type: S   Sect: 6 Obj: 4   RANDOM: Y

Using the t-table, find the critical value for the given situation.

34. $n = 8$, $\alpha = 0.005$, one-tailed left test

Answer: -3.499 Type: S   Sect: 7 Obj: 6   RANDOM: Y

Use the steps in hypothesis testing to solve the problem.

35. A grocer claims that the average grocery bill at his store is $95. A sample of 20 customers' bills showed an average of $98 with a standard deviation of $7.00. At $\alpha = 0.05$, is there evidence to support the grocer's claim?

Answer:
$H_0 : \mu = 95$; $H_1 : \mu \neq 95$
C.V. $= \pm 2.093$, d.f. $= 19$
$t = 1.92$
Do not reject H_0; the mean is $95.

Type: S   Sect: 7 Obj: 7   RANDOM: Y

36. A restaurant in El Paso claims that it will serve lunch to a customer in 12 minutes or less or the customer will get lunch free. In order to test the claim, 25 customers kept track of how long it took to get their meal served. Their average time was 13.2 minutes with a standard deviation of 1.5 minutes. At $\alpha = 0.01$, should the restaurant change its claim?

Answer:
$H_0 : \mu \leq 12$; $H_1 : \mu > 12$
C.V. $= 2.492$, d.f. $= 24$
Reject H_0; the restaurant takes longer than 12 minutes to serve customers lunch.

Type: S   Sect: 7 Obj: 7   RANDOM: Y

37. The average water bill in Quail Heights is believed to be $15.00 per month. One subdivision felt that its water bills were higher than the average, so it took a sample of 27 homes in the subdivision and found the average water bill to be $16.50 with a standard deviation of $2.25. Is there evidence to support the subdivision's claim? Let $\alpha = 0.05$.

Answer:
$H_0 : \mu \leq 15$; $H_1 : \mu > 15$
C.V. $= 1.706$, d.f. $= 26$
$t = 3.46$
Reject H_0; the subdivision has higher-than-average water bills.

Type: S   Sect: 7 Obj: 7   RANDOM: Y

38. A survey showed that the average price of a movie ticket in Norman was $5.00. A sample of 22 theaters showed the average price of their tickets was $5.30 with a standard deviation of $0.75. Is the survey correct? Use $\alpha = 0.05$.

Answer:
$H_0 : \mu = 5$; $H_1 : \mu \neq 5$
C.V. $= \pm 2.080$, d.f. $= 21$
$t = 1.88$
Do not reject H_0; the average price of tickets is $5.00.

Type: S   Sect: 7 Obj: 7   RANDOM: Y

39. A class of 25 seventh graders has completed a standardized math test on which their average was 45 and their standard deviation was 7.5. The national average on the math test is 52. Do we have sufficient evidence to claim that this seventh grade class is inferior in mathematical ability? Use $\alpha = 0.01$.

Answer:
$H_0 : \mu \geq 52$; $H_1 : \mu < 52$
C.V. $= -2.492$, d.f. $= 24$
$t = -4.67$
Reject H_0; this class of seventh graders appears to be inferior in mathematical ability.

Type: S   Sect: 7 Obj: 7   RANDOM: Y

40. The average daily wage in a particular industry is $80. A factory at Bridgeport feels that it is making below the average, so it randomly selected 20 employees and found that their average daily wage was $76 with a standard deviation of $6. At $\alpha = 0.025$, is there evidence to support the factory's claim?

Answer:
$H_0 : \mu \geq 80$; $H_1 : \mu < 80$
C.V. $= -2.093$, d.f. $= 19$
$t = -2.98$
Reject H_0; there is evidence to show that this factory has lower wages.

Type: S   Sect: 7 Obj: 7   RANDOM: Y

41. An automobile dealer believes that his new model will give trouble-free service for at least 15,000 miles. In a test of 8 new models, the following numbers of trouble-free miles were recorded: 16,000, 15,000, 14,350, 14,500, 13,750, 13,000, 14,250, 13,900. Do these data refute the automobile dealer's claim? Let $\alpha = 0.025$.

    Answer:
    $H_0 : \mu \geq 15,000$; $H_1 : \mu < 15,000$
    C.V. = -2.365, d.f. = 7
    $t = -2.08$
    Do not reject H_0. It appears that the dealer is correct in his statement that his new model will give trouble-free service for at least 15,000 miles.

    Type: S   Sect: 7 Obj: 7   RANDOM: Y

42. A stenography school claims that its graduating students will average 80 words per minute in dictation. A sample of 18 graduates found that they averaged 78 words per minute in dictation with a standard deviation of 8 words per minute. At $\alpha = 0.05$, is the school's claim correct?

    Answer:
    $H_0 : \mu = 80$; $H_1 : \mu \neq 80$
    C.V. = ±2.110, d.f. = 17
    $t = -1.06$
    Do not reject H_0; the school's claim is valid.

    Type: S   Sect: 7 Obj: 7   RANDOM: Y

43. A biology teacher claims that the average on her last test was 74. The scores of 12 of her students are as follows: 78, 82, 90, 65, 81, 66, 72, 55, 58, 70, 74, 85. At the 0.01 level of significance, is her claim correct?

    Answer:
    $H_0 : \mu = 74$; $H_1 : \mu \neq 74$
    C.V. = ±3.106, d.f. = 11
    $t = -0.32$
    Do not reject H_0; the teacher's claim is correct.

    Type: S   Sect: 7 Obj: 7   RANDOM: Y

44. A burger chain advertises that it puts at least 0.33 pounds of beef in its hamburgers. Jerry, who eats at this chain fairly often, does not believe this claim. He randomly selects 10 hamburgers and determines that they average 0.31 pounds of beef with a standard deviation of 0.08 pound. Is the restaurant's claim correct, if $\alpha = 0.025$?

Answer:
$H_0 : \mu \geq 0.33$; $H_1 : \mu < 0.33$
C.V. = -2.262, d.f. = 9
$t = -0.79$
Do not reject H_0; the restaurant's claim appears to be correct.

Type: S   Sect: 7 Obj: 7   RANDOM: Y

45. An inspector from the Department of Weights and Measures weighs 12 eighteen-ounce cereal boxes of corn flakes. He finds their mean weight to be 17.85 ounces with a standard deviation of 0.3 ounce. Are the cereal boxes lighter than they should be? Let $\alpha = 0.01$.

Answer:
$H_0 : \mu \geq 18$; $H_1 : \mu < 18$
C.V. = -2.718, d.f. = 11
$t = -1.73$
Do not reject H_0. The cereal boxes are not lighter than they should be.

Type: S   Sect: 7 Obj: 7   RANDOM: Y

46. A jewelry store claims that the average age of its customers is over 38. A random sample of 25 customers yielded an average age of 45 with a standard deviation of 4 years. At $\alpha = 0.01$, can the jewelry store's claim be substantiated?

Answer:
$H_0 : \mu \leq 38$; $H_1 : \mu > 38$
C.V. = 2.492, d.f. = 24
$t = 8.75$
Reject H_0; the jewelry store's claim is correct.

Type: S   Sect: 7 Obj: 7   RANDOM: Y

47. The average price of steaks in Homer is thought to be $7.50. Bill's steak house claims to have the lowest price of steaks in Homer. The average price of 10 of his steaks is $6.50 with a standard deviation of $1.25. At $\alpha = 0.01$, can we say that Bill is right?

Answer:
$H_0 : \mu \geq 7.50$; $H_1 : \mu < 7.50$
C.V. = -2.821, d.f. = 9
$t = -2.53$
Do not reject H_0. We cannot conclude that Bill is right.

Type: S   Sect: 7 Obj: 7   RANDOM: Y

Use the steps in hypothesis testing to solve the problem.

48. It has been claimed that at most 12% of the restaurants in Wichita specialize in hamburgers. A sample of 150 restaurants shows that 23 specialize in hamburgers. Check this claim at $\alpha = 0.025$.

    Answer:
    $H_0 : p \leq 0.12$; $H_1 : p > 0.12$
    C.V. = 1.96; $z = 1.26$
    Do not reject H_0. It appears that the claim that at most 12% of the restaurants in Wichita specialize in hamburgers is correct.

    Type: S   Sect: 8 Obj: 8   RANDOM: Y

49. It has been claimed that at least 35% of the students at OCU live on campus. From a sample of 300 students, 98 live on campus. Does the evidence support this claim at $\alpha = 0.01$?

    Answer:
    $H_0 : p \geq 0.35$; $H_1 : p < 0.35$
    C.V. = -2.33; $z = -0.85$
    Do not reject H_0. The claim that at least 35% of the students at OCU live on campus is correct.

    Type: S   Sect: 8 Obj: 8   RANDOM: Y

50. A survey in Middleton revealed that at least 55% of the mothers with school-age children worked outside the home. A sample of 150 mothers revealed that 96 were working outside the home. Using $\alpha = 0.05$, is the survey correct?

    Answer:
    $H_0 : p \leq 0.55$; $H_1 : p > 0.55$
    C.V. = 1.645; $z = 2.22$
    Reject H_0. It appears that more than 55% of Middleton mothers with school-age children work outside the home.

    Type: S   Sect: 8 Obj: 8   RANDOM: Y

51. A recent survey in the state of Texas stated that 23% of drivers on the freeways were carrying a gun. An anonymous questionnaire was sent out to 200 Texans, and 50 responded that they carry a gun in their car. Is there evidence to support the survey? Let $\alpha = 0.01$.

Answer:
$H_0 : p = 0.23;\ H_1 : p \neq 0.23$
C.V. $= \pm 2.575;\ z = 0.67$
Do not reject H_0. The claim is supported.

Type: S   Sect: 8   Obj: 8   RANDOM: Y

52. A study done at Bridge Creek reported that no more than 15% of the people living there made less than \$20,000 a year. To check this claim, a statistician polled 90 people in Bridge Creek and found that 10 made less than \$20,000 a year. Was the study's report accurate? Let $\alpha = 0.025$.

Answer:
$H_0 : p \geq 0.15;\ H_1 : p < 0.15$
C.V. $= -1.96;\ z = -1.03$
Do not reject H_0. The study's report is accurate.

Type: S   Sect: 8   Obj: 8   RANDOM: Y

53. A random sample of 500 shoppers at Quincy Mall found that 145 favored longer shopping hours. Is this sufficient evidence to conclude that less than 33% of the shoppers at Quincy Mall favor longer hours? Let $\alpha = 0.05$.

Answer:
$H_0 : p \geq 0.33;\ H_1 : p < 0.33$
C.V. $= -1.645;\ z = -1.90$
Reject H_0. It appears that less than 33% of shoppers prefer longer hours.

Type: S   Sect: 8   Obj: 8   RANDOM: Y

54. Warren Porter is running for mayor. He thinks that at least 60% of the voters will vote for him. A random survey of 150 voters showed that 102 people would vote for Warren. Is there evidence to support Warren's claim at $\alpha = 0.025$?

Answer:
$H_0 : p \leq 0.6;\ H_1 : p > 0.6$
C.V. $= 1.96;\ z = 2$
Reject H_0. Warren's claim is correct. It appears that at least 60% of the voters will vote for him.

Type: S   Sect: 8   Obj: 8   RANDOM: Y

55. A small college claimed that no more than 20% of its students commute more than 10 miles. A random sample of 100 students showed that 24 students traveled more than 10 miles to college. At $\alpha = 0.05$, can we conclude that the college's claim is incorrect?

Answer:
$H_0 : p \leq 0.2;\ H_1 : p > 0.2$
C.V. $= 1.645;\ z = 1$
Do not reject H_0. The college's claim appears to be correct.

Type: S   Sect: 8 Obj: 8   RANDOM: Y

56. A survey stated that more than 40% of people starting diets in January get discouraged and quit dieting within 2 weeks. To test this claim, a researcher asked 400 people who had started diets in early January if they were still dieting two weeks later. The results were that 185 were no longer dieting. Can we conclude that the survey was correct? Let $\alpha = 0.01$.

Answer:
$H_0 : p \leq 0.4;\ H_1 : p > 0.4$
C.V. $= 2.33;\ z = 2.55$
Reject H_0. The survey is correct. It appears that more than 40% of people starting diets in January get discouraged and quit within 2 weeks.

Type: S   Sect: 8 Obj: 8   RANDOM: Y

57. A claim was made that the average score on the ACT test for students at Jefferson College was 19 with a standard deviation of 3. A sample of 81 students had an average score of 17.8. What is the null hypothesis for these data?

Answer: $H_0 : \mu = 19$ Type: S   Sect: 3 Obj: 3   RANDOM: Y

58. A claim was made that the average score on the ACT test for students at Jefferson College was 19 with a standard deviation of 3. A sample of 81 students had an average score of 17.8. What is the alternative hypothesis for these data?

Answer: $H_1 : \mu \neq 19$ Type: S   Sect: 3 Obj: 3   RANDOM: Y

59. A claim was made that the average score on the ACT test for students at Jefferson College was 19 with a standard deviation of 3. A sample of 81 students had an average score of 17.8. If $\alpha = 0.01$, what is the critical value for these data?

Answer: ± 2.575 Type: S   Sect: 6 Obj: 3   RANDOM: Y

60. A claim was made that the average score on the ACT test for students at Jefferson College was 19 with a standard deviation of 3. A sample of 81 students had an average score of 17.8. If $\alpha = 0.01$, what is the test statistic for these data?

Answer: -3.6 Type: S   Sect: 6 Obj: 3   RANDOM: Y

61. A claim was made that the average score on the ACT test for students at Jefferson College was 19 with a standard deviation of 3. A sample of 81 students had an average score of 17.8. If $\alpha = 0.01$, what conclusion can be made?

Answer: Reject H_0. The average ACT score at Jefferson College was not 19. Type: S

Sect: 6 Obj: 3   RANDOM: Y

62. A child psychologist claims that the average number of mistakes an eight-year-old will make on a skills assessment test is 6 with a standard deviation of 3. A random sample of 49 eight-year-olds had a mean of 7 mistakes. Test the psychologist's claim at $\alpha = 0.025$. What is the P-value?

Answer: 0.0198 Type: S   Sect: 11 Obj: 9   RANDOM: Y

63. A calculus student claims that the average calculus textbook costs at least $60. The standard deviation is $4.00. A sample of 36 books has an average cost of $62. At $\alpha = 0.01$, test the student's claim. What is the P-value?

Answer: 0.0013 Type: S   Sect: 11 Obj: 9   RANDOM: Y

64. The manager of a pizza parlor claims his employees make an average of $4.50 per hour with a standard deviation of $0.75. Bob does not believe his claim, so he takes a sample of 40 employees and find their average salary is $4.25. Test the manager's claim at $\alpha = 0.05$. What is the P-value?

Answer: 0.0348 Type: S   Sect: 11 Obj: 9   RANDOM: Y

65. To justify raising car insurance rates, an insurance company claims that the mean accident expense is at least $1,250 per year with a standard deviation of $125. In a survey of 100 randomly selected accident reports, it was found that the mean accident expense was $1,230. At $\alpha = 0.025$, test the claim that the insurance company is misinformed. What is the P-value?

Answer: 0.0548 Type: S   Sect: 11 Obj: 9   RANDOM: Y

CHAPTER 10

TRUE/FALSE

1. One assumption that must be met in order to test the difference between two independent means is that the populations from which the samples were obtained must be normally distributed, and the standard deviations of the variable must be known.

   Answer: T Type: T   Sect: 2 Obj: 1   RANDOM: Y

2. A research firm tested two different makes of pickups to determine if there was a difference in the average number of miles per gallon each make obtained. They tested 20 models of make 1 and the average was 23 miles per gallon with a standard deviation of 3. Fifteen models of make 2 had an average of 25 miles per gallon with a standard deviation of 2. The test statistic for this hypothesis test is 2.15.

   Answer: F Type: T   Sect: 2 Obj: 1   RANDOM: Y

3. When testing the difference between two means using the t-test for independent samples, the variances are assumed to be equal. Sample 1 has 8 items in it and sample 2 has 9 items in it. The number of degrees of freedom for this test would be 15.

   Answer: T Type: T   Sect: 3 Obj: 2   RANDOM: Y

4. A two-tailed t-test was used for testing the difference between two independent means. The variances were assumed to be equal. Sample A had 12 people in it, and sample B had 9 people in it. If $\alpha = 0.05$, the critical value would be 2.201.

   Answer: F Type: T   Sect: 3 Obj: 2   RANDOM: Y

5. A psychologist wants to see if a 2-hour course on how to improve your memory will actually do what is claimed. Sixteen people are given a memory test both before and after the 2-hour course. This would be an example of dependent samples.

   Answer: T Type: T   Sect: 4 Obj: 3   RANDOM: Y

6. Two different brands of cigarettes were tested for nicotine content. These samples would be dependent samples.

   Answer: F Type: T   Sect: 4 Obj: 3   RANDOM: Y

7. Twelve men were given a test to determine how many pounds of weights they could lift before and after taking steroids. The differences were as follows: -3, +12, +15, -1, +30, +14, 0, +40, +6, -2, +19, +24. The mean of the differences would be equal to 12.83.

   Answer: T Type: T   Sect: 4 Obj: 4   RANDOM: Y

8. When testing small independent samples where the population variances are equal, sample 1 had 12 items and sample 2 contained 10 items. The number of degrees of freedom for this hypothesis test would be 21.

   Answer: F Type: T   Sect: 4 Obj: 4   RANDOM: Y

9. If we say that there is no difference between two population means, then we would be conducting a two-tailed hypothesis test.

   Answer: T Type: T   Sect: 1 Obj: 1   RANDOM: Y

10. Two samples are independent when knowledge of the observations in one sample provide no information about observations in the other sample.

    Answer: T Type: T   Sect: 2 Obj: 1   RANDOM: Y

MULTIPLE CHOICE

11. In testing the difference between two independent means using the z-test, if both populations have the same mean, then most of the differences should be close to what number?
    a. 1
    b. 2
    c. 0
    d. -1

    Answer: c Type: M   Sect: 2 Obj: 1   RANDOM: Y

12. A student claimed that the average grade in the night section of statistics was higher than the average grade in the day section. She used a sample of 25 students from the night section and calculated a mean of 80 with a standard deviation of 10. A sample of 20 students from the day section had a mean of 78 and a standard deviation of 13. If she is using a z-test to compare independent samples, what is the test statistic?
    a. 0.567
    b. 0.408
    c. 0.583
    d. 0.402

    Answer: a Type: M   Sect: 2 Obj: 1   RANDOM: Y

13. Which of the following is a reason for using the t-test for independent samples when comparing two means?
    a. when sampling from two binomial populations
    b. when the standard deviations are not known and the sample size is smaller than 31
    c. when the sample size is greater than or equal to 31
    d. when the standard deviations of the population are known

    Answer: b Type: M   Sect: 3 Obj: 2   RANDOM: Y

14. A one-tailed right t-test for independent samples was performed. The variances are assumed to be equal. One sample asked opinions of 14 children and a second sample asked opinions of 17 children. If
$\alpha = 0.01$, what is the critical value for this test?
    a. 2.65
    b. 2.583
    c. 2.462
    d. 2.326

Answer: c Type: M   Sect: 3 Obj: 2   RANDOM: Y

15. Which of the following is not an example of dependent samples?
    a. effects of a drug on reaction time measured by a before-and-after test
    b. test scores of the same students in math and chemistry
    c. the effectiveness of a reading improvement course by using a
       pre-test and a post-test on the people enrolled
    d. the blood pressure readings of overweight males and overweight females

Answer: d Type: M   Sect: 4 Obj: 3   RANDOM: Y

16. Listed below is a table of before-and-after quiz scores.

Before	5	8	9	6	4	4	5	8	10	6	5
After	8	7	9	9	8	6	8	9	14	9	6

What is the standard deviation of the differences?
    a. 2.69
    b. 1.64
    c. 2.73
    d. 1.61

Answer: b Type: M   Sect: 4 Obj: 4   RANDOM: Y

17. Listed below is a table of before-and-after quiz scores.

Before	5	8	9	6	4	4	5	8	10	6	5
After	8	7	9	9	8	6	8	9	14	9	6

What is the test value?
    a. -4.23
    b. -4.59
    c. -2.58
    d. -4.03

Answer: a Type: M   Sect: 4 Obj: 4   RANDOM: Y

18. Using the t-test for $n_1 = 12$, $s_1 = 5$, $n_2 = 10$, $s_2 = 4$, and assuming variances are equal, what is the value of the pooled estimator of the population variance?
    a. 4.58   b. 20.95   c. 3.84   d. 2.05

Answer: b Type: M   Sect: 3 Obj: 4   RANDOM: Y

19. Six randomly selected students complete a quiz before learning about fractions and another quiz after learning about fractions. The results are listed below.

Student	Scores	
	Before	After
1	15	16
2	8	12
3	10	14
4	9	10
5	17	17
6	12	11

What is the mean of the differences?
a. –9   b. -2.5   c. -0.67   d. -1.5

Answer: d Type: M   Sect: 4 Obj: 4   RANDOM: Y

20. Six randomly selected students complete a quiz before learning about fractions and another quiz after learning about fractions. The results are listed below.

Student	Scores	
	Before	After
1	15	16
2	8	12
3	10	14
4	9	10
5	17	17
6	12	11

What is the standard error of the $\overline{D}$ values?
a. 0.85   b. 1.76   c. 0.72   d. 0.51

Answer: a Type: M   Sect: 4 Obj: 4   RANDOM: Y

SHORT ANSWER

Use the steps in hypothesis testing to solve the problem.

21. A nutritionist wishes to compare the number of calories in hamburgers at 2 different national chains to determine if there is a difference between them. The average number of calories of each are listed below. Test the claim at $\alpha = 0.01$. (Use the z-test.)

<u>Chain 1</u>	<u>Chain 2</u>
$\bar{x}_1 = 485$	$\bar{x}_2 = 472$
$\sigma_1 = 4$	$\sigma_2 = 3.5$
$n_1 = 25$	$n_2 = 25$

Answer:
$H_0 : \mu_1 - \mu_2 = 0;\ H_1 : \mu_1 - \mu_2 \neq 0$
C.V. $= \pm 2.595$
$z = 12.23$
Reject H_0. There is a difference between the two chains in the number of calories their hamburgers contain.

Type: S   Sect: 2 Obj: 1   RANDOM: Y

22. The head of the Engineering Department at Sumner College thinks his graduates have a higher grade-point average than those at Compton College. A sample of 100 graduates is used and their average GPA is calculated. Test his claim at $\alpha = 0.025$.

<u>Sumner College</u>	<u>Compton College</u>
$\bar{x}_1 = 3.2$	$\bar{x}_2 = 3.15$
$\sigma_1 = 0.75$	$\sigma_2 = 0.62$
$n_1 = 100$	$n_2 = 100$

Answer:
$H_0 : \mu_1 - \mu_2 \leq 0;\ H_1 : \mu_1 - \mu_2 > 0$
C.V. $= 1.96$
$z = 0.51$
Do not reject H_0. It does not appear that the graduates of Sumner College have a higher grade-point average than those of Compton College.

Type: S   Sect: 2 Obj: 1   RANDOM: Y

23. Joan wants to purchase a new couch. She wants to determine if there is any difference between the average cost of couches at 2 different stores. Test the hypothesis that there is no difference at $\alpha = 0.05$.

Store 1	Store 2
$\bar{x}_1 = \$680$	$\bar{x}_2 = \$720$
$\sigma_1 = \$\ 65$	$\sigma_2 = \$\ 75$
$n_1 = 35$	$n_2 = 32$

Answer:

$H_0 : \mu_1 - \mu_2 = 0;\ H_1 : \mu_1 - \mu_2 \neq 0$

C.V. $= \pm1.96$

$z = -2.32$

Reject H_0. There is a difference in price between the two stores.

Type: S   Sect: 2 Obj: 1   RANDOM: Y

24. A doctor wanted to determine if the average number of minutes spent in physical activity per day was less for his overweight patients than for his other patients. Check his hypothesis at $\alpha = 0.025$.

Overweight Patients	Other Patients
$\bar{x}_1 = 15$ min.	$\bar{x}_2 = 40$ min.
$\sigma_1 = \ 3$ min.	$\sigma_2 = \ 5$ min.
$n_1 = 35$	$n_2 = 40$

Answer:

$H_0 : \mu_1 - \mu_2 \geq 0;\ H_0 : \mu_1 - \mu_2 < 0$

C.V. $= -1.96$

$z = -26.62$

Reject H_0. The doctor is correct in his hypothesis that his overweight patients spend less time in physical activity than his other patients.

Type: S   Sect: 2 Obj: 1   RANDOM: Y

25. A researcher wanted to know if the average number of phone calls received per day was the same in the city as in the country. Her hypothesis was there would be no difference in the number of calls. Test her claim at $\alpha = 0.05$.

Country	City
$\overline{x}_1 = 12$	$\overline{x}_2 = 10$
$\sigma_1 = 3$	$\sigma_2 = 2$
$n_1 = 50$	$n_2 = 50$

Answer:
$H_0 : \mu_1 - \mu_2 = 0$; $H_1 : \mu_1 - \mu_2 \neq 0$
C.V. $= \pm 1.96$
$z = 3.92$
Reject H_0. There is a difference between the city and the country in the number of phone calls received.

Type: S   Sect: 2 Obj: 1   RANDOM: Y

26. A conservationist wants to know whether the average water level of Horseshoe Lake is higher than the average water level of Swan Lake. Test his hypothesis at $\alpha = 0.01$. (He took a number of readings at each lake.)

Horseshoe Lake	Swan Lake
$\overline{x}_1 = 38$ feet	$\overline{x}_2 = 36$ feet
$\sigma_1 = 2$ feet	$\sigma_2 = 1.5$ feet
$n_1 = 35$	$n_2 = 35$

Answer:
$H_0 : \mu_1 - \mu_2 \leq 0$; $H_1 : \mu_1 - \mu_2 > 0$
C.V. $= 2.33$
$z = 4.73$
Reject H_0. It appears that the average water level of Horseshoe Lake is higher than the average water level of Swan Lake.

Type: S   Sect: 2 Obj: 1   RANDOM: Y

27. It has been hypothesized that women in the city spend less time preparing meals than women in the country. To test this hypothesis a sample of 40 women were chosen and their average preparation times were recorded. Test this hypothesis at $\alpha = 0.05$.

<u>City</u>	<u>Country</u>
$\overline{x}_1 = 20$ min.	$\overline{x}_2 = 30$ min.
$\sigma_1 = 3$ min.	$\sigma_2 = 5$ min.
$n_1 = 40$	$n_2 = 40$

Answer:
$H_0 : \mu_1 - \mu_2 \geq 0$; $H_1 : \mu_1 - \mu_2 < 0$
C.V. -1.645
$z = -10.85$
Reject H_0. The hypothesis that women in the city spend less time preparing meals than women in the country is correct.

Type: S   Sect: 2 Obj: 1   RANDOM: Y

28. A consumer advocate hypothesizes that there is no difference between two different brands of televisions in the average lifetime of their picture tubes. Use $\alpha = 0.05$ to test his claim.

<u>Brand A</u>	<u>Brand B</u>
$\overline{x}_1 = 6.5$ yrs.	$\overline{x}_2 = 7.7$ yrs.
$\sigma_1 = 0.7$ yr	$\sigma_2 = 0.9$ yr
$n_1 = 45$	$n_2 = 50$

Answer:
$H_0 : \mu_A - \mu_B = 0$; $H_1 : \mu_A - \mu_B \neq 0$
C.V. $= \pm 1.96$
$z = -7.29$
Reject H_0. There is a difference between the two brands in the average lifetime of their picture tubes.

Type: S   Sect: 2 Obj: 1   RANDOM: Y

29. A company has two branches, one on the East Coast and the other on the West Coast. The president of the company hypothesizes that there is no difference in the average daily sales of the two branches. He got the sales figures for each branch from the last 60 days. Test his hypothesis at $\alpha = 0.01$.

<table>
<tr><td><u>East Coast Branch</u></td><td><u>West Coast Branch</u></td></tr>
<tr><td>$\overline{x}_1 = \$18,000$</td><td>$\overline{x}_2 = \$19,500$</td></tr>
<tr><td>$\sigma_1 = \$\ 1,700$</td><td>$\sigma_2 = \$\ 2,300$</td></tr>
<tr><td>$n_1 = 60$</td><td>$n_2 = 60$</td></tr>
</table>

Answer:
$H_0 : \mu_1 - \mu_2 = 0$; $H_1 : \mu_1 - \mu_2 \neq 0$
C.V. $= \pm 2.575$
$z = -4.06$
Reject H_0. There is a difference in average sales between the two branches of this company.

Type: S   Sect: 2 Obj: 1   RANDOM: Y

30. A consumer advocate hypothesizes that there is no difference between two different brands of televisions in the average lifetime of their picture tubes. Use $\alpha = 0.05$ to test his claim.

<table>
<tr><td><u>Brand A</u></td><td><u>Brand B</u></td></tr>
<tr><td>$\overline{x}_1 = 6.5$ yrs.</td><td>$\overline{x}_2 = 6.9$ yrs.</td></tr>
<tr><td>$\sigma_1 = 0.7$ yr</td><td>$\sigma_2 = 0.9$ yr</td></tr>
<tr><td>$n_1 = 42$</td><td>$n_2 = 48$</td></tr>
</table>

Answer:
$H_0 : \mu_A - \mu_B = 0$; $H_1 : \mu_A - \mu_B \neq 0$
C.V. $= \pm 1.96$
$z = -2.37$
Reject H_0. There is a difference between the two brands in the average lifetime of their picture tubes.

Type: S   Sect: 2 Obj: 1   RANDOM: Y

31. A study was conducted to compare the speed limits in two different areas of town to see if there was any difference. Use $\alpha = 0.05$ to test for a difference.

<u>Area 1</u>   <u>Area 2</u>

$\bar{x}_1 = 35$ mph $\bar{x}_2 = 25$ mph
$\sigma_1 = 3$ mph $\sigma_2 = 2$ mph
$n_1 = 44$ $n_2 = 40$

Answer:
$H_0 : \mu_1 - \mu_2 = 0$; $H_1 : \mu_1 - \mu_2 \neq 0$
C.V. $= \pm 1.96$
$z = 18.12$
Reject H_0. There is a difference in speed limits between the two different areas of town.

Type: S   Sect: 2   Obj: 1   RANDOM: Y

32. Two methods of teaching first graders to read are being compared. The data are listed below.

<u>Method 1</u>   <u>Method 2</u>

$\bar{x}_1 = 65.7$ $\bar{x}_2 = 72.5$
$\sigma_1 = 6.3$ $\sigma_2 = 5.6$
$n_1 = 45$ $n_2 = 50$

Can we conclude that there is no difference in the two methods at $\alpha = 0.02$?

Answer:
$H_0 : \mu_1 - \mu_2 = 0$; $H_1 : \mu_1 - \mu_2 \neq 0$
C.V. $= \pm 2.33$
$z = -5.54$
Reject H_0. There is a difference in the two methods of teaching.

Type: S   Sect: 2   Obj: 1   RANDOM: Y

33. A researcher wants to determine if the salaries of engineers working for the State Department of Transportation are lower than those of engineers working for private industry. She selects a sample of engineers from each group and calculates their mean and standard deviations. At $\alpha = 0.025$, can it be concluded that state employees earn less than employees in private industry? (Variances are equal.)

State	Private Industry
$\bar{x}_1 = \$32,450$	$\bar{x}_2 = \$36,000$
$s_1 = \$\ \ \ 900$	$s_2 = \$\ 1,200$
$n_1 = 15$	$n_2 = 14$

Answer:
$H_0 : \mu_s - \mu_p \geq 0$; $H_1 : \mu_s - \mu_p < 0$
C.V. $= -2.052$, d.f. $= 27$
$t = -9.06$
Reject H_0. The state employees do earn less than those in private industry.

Type: S   Sect: 3   Obj: 2   RANDOM: Y

34. A local charity thinks that people in River Heights give more money to their charity than people in Lakeview. It conducted a survey of 25 people in each subdivision and recorded the results. Is its hypothesis correct? Let $\alpha = 0.01$. Assume variances are equal.

River Heights	Lakeview
$\bar{x}_1 = \$30$	$\bar{x}_2 = \$20$
$s_1 = \$\ 3$	$s_2 = \$\ 4$
$n_1 = 25$	$n_2 = 25$

Answer:
$H_0 : \mu_1 - \mu_2 \leq 0$; $H_1 : \mu_1 - \mu_2 > 0$
C.V. $= 2.33$
$t = 10$
Reject H_0. It appears that the people in River Heights do give more money to the local charity than the people in Lakeview.

Type: S   Sect: 3   Obj: 2   RANDOM: Y

35. An agronomist felt that corn-fed hogs would have an average weight that was greater than that of pellet-fed hogs. He conducted a study to determine if his theory was correct. Check his claim at $\alpha = 0.01$. (Variances are equal.)

Corn-fed Hogs	Pellet-fed Hogs
$\bar{x}_1 = 280$ lbs.	$\bar{x}_2 = 270$ lbs.
$s_1 = 10$ lbs.	$s_2 = 15$ lbs.
$n_1 = 15$	$n_2 = 13$

Answer:
$H_0 : \mu_1 - \mu_2 \leq 0;\ H_1 : \mu_1 - \mu_2 > 0$
C.V. $= 2.479$, d.f. $= 26$
$t = 2.10$
Do not reject H_0. There is not enough evidence to show that corn-fed hogs have an average weight that is greater than that of pellet-fed hogs.

Type: S   Sect: 3 Obj: 2   RANDOM: Y

36. A shopper felt that there was no difference between the price of Eveready batteries and DuraCell batteries. To test her claim she checked the prices on batteries at 10 different stores. Is her claim correct at $\alpha = 0.10$? (Variances are equal.)

Eveready	DuraCell
$\bar{x}_1 = \$2.14$	$\bar{x}_2 = \$2.08$
$s_1 = \$0.10$	$s_2 = \$0.11$
$n_1 = 10$	$n_2 = 10$

Answer:
$H_0 : \mu_1 - \mu_2 = 0;\ H_1 : \mu_1 - \mu_2 \neq 0$
C.V. $= \pm 1.734$, d.f. $= 18$
$t = 1.28$
Do not reject H_0. There does not appear to be a difference in price between the two types of batteries.

Type: S   Sect: 3 Obj: 2   RANDOM: Y

37. A discount food chain claims that it has the lowest prices on toilet paper. Cathy decides to check their claim by finding prices of 8 different brands of toilet paper at the discount store and at another supermarket in town. At $\alpha = 0.01$, is the discount store's claim correct? (Assume variances are equal.)

Discount Store	Supermarket X
$\overline{x}_1 = \$1.09$	$\overline{x}_2 = \$1.25$
$s_1 = \$0.10$	$s_2 = \$0.08$
$n_1 = 8$	$n_2 = 8$

Answer:
$H_0 : \mu_1 - \mu_2 \geq 0$; $H_1 : \mu_1 - \mu_2 < 0$
C.V. $= -2.624$, d.f. $= 14$
$t = -3.53$
Reject H_0. The discount food chain does seem to have the lowest prices on toilet paper.

Type: S   Sect: 3   Obj: 2   RANDOM: Y

38. A fisherman claims that the average length of catfish caught at Bear Lake is greater than the average length of catfish caught at Crystal Lake. To check his claim he went fishing at both lakes and the data are listed below. At $\alpha = 0.025$, is the fisherman correct? (Assume variances are equal.)

Bear Lake	Crystal Lake
$\overline{x}_1 = 15.5$ in.	$\overline{x}_2 = 14.25$ in.
$s_1 = 1.25$ in.	$s_2 = 1.00$ in.
$n_1 = 14$	$n_2 = 12$

Answer:
$H_0 : \mu_B - \mu_C \leq 0$; $H_1 : \mu_B - \mu_C > 0$
C.V. $= 2.064$, d.f. $= 24$
$t = 2.78$
Reject H_0. There seems to be evidence to show that catfish at Bear Lake have an average length greater than the length of catfish caught at Crystal Lake.

Type: S   Sect: 3   Obj: 2   RANDOM: Y

39. A researcher believes that students in the Eastern schools have a better average on a standardized test in mathematics than students in Southern schools. To test this claim, 20 students were randomly selected, and their mean and standard deviation were calculated. Is the researcher correct in his claim? Use $\alpha = 0.025$ and assume that variances are equal.

Eastern Schools   Southern Schools

$\overline{x}_1 = 82.4$ $\overline{x}_2 = 79.3$

$s_1 = 5.2$ $s_2 = 4.8$

$n_1 = 20$ $n_2 = 18$

Answer:

$H_0 : \mu_1 - \mu_2 \le 0$; $H_1 : \mu_1 - \mu_2 > 0$

C.V. $= 1.96$

$t = 1.90$

Do not reject H_0. We cannot conclude that Eastern schools have a better average on math test than Southern schools.

Type: S   Sect: 3 Obj: 2   RANDOM: Y

40. Donaldson Corporation wants to hire a temporary secretary. There are 2 employment agencies in town and it is believed that the average hourly wages charged by both agencies are the same. Test this claim at $\alpha = 0.05$. Assume that variances are equal.

Agency A   Agency B

$\overline{x}_1 = \$6.50$ $\overline{x}_2 = \$6.75$

$s_1 = \$0.50$ $s_2 = \$0.65$

$n_1 = 18$ $n_2 = 20$

Answer:

$H_0 : \mu_A - \mu_B = 0$; $H_1 : \mu_A - \mu_B \ne 0$

C.V. $= \pm 1.96$

$t = -1.32$

Do not reject H_0. There does not appear to be a difference in wages between the two agencies.

Type: S   Sect: 3 Obj: 2   RANDOM: Y

41. A doctor wanted to determine if the average number of minutes spent in physical activity per day was less for his overweight patients than for his other patients. Check his hypothesis at $\alpha = 0.025$. Variances are assumed to be equal.

Overweight Patients	Other Patients
$\bar{x}_1 = 20$ min.	$\bar{x}_2 = 45$ min.
$s_1 = 3$ min.	$s_2 = 7$ min.
$n_1 = 24$	$n_2 = 18$

Answer:

$H_0 : \mu_1 - \mu_2 \geq 0$; $H_1 : \mu_1 - \mu_2 < 0$

C.V. $= -1.96$

$t = -15.72$

Reject H_0. The doctor's claim that his overweight patients exercise less than his other patients seems to be correct.

Type: S   Sect: 3 Obj: 2   RANDOM: Y

42. A director of intramural sports wanted to see if there was a difference in height between intramural basketball players and varsity players on the university team. Variances are assumed to be equal. Check the hypothesis that the average height of varsity players is greater than the average height of intramural players at $\alpha = 0.01$. The heights are in inches.

Intramural Players	Varsity Players
$\bar{x}_1 = 76$ in.	$\bar{x}_2 = 80$ in.
$s_1 = 3$ in.	$s_2 = 2$ in.
$n_1 = 16$	$n_2 = 15$

Answer:

$H_0 : \mu_V - \mu_I \leq 0$; $H_1 : \mu_V - \mu_I > 0$

C.V. $= 2.462$, d.f. $= 29$

$t = 4.34$

Reject H_0. The height of varsity players is greater than the height of intramural players.

Type: S   Sect: 3 Obj: 2   RANDOM: Y

43. A pharmacist wants to check the average time it takes for 2 different pain relievers to alleviate a tension headache. He thinks that there is no difference between the two pain relievers. Test his claim at $\alpha = 0.05$. Variances are assumed to be equal.

<table>
<tr><td colspan="2">Pain Reliever 1</td><td colspan="2">Pain Reliever 2</td></tr>
<tr><td>$\bar{x}_1$ = 15 min.</td><td></td><td>$\bar{x}_2$ = 12 min.</td><td></td></tr>
<tr><td>s_1 = 3 min.</td><td></td><td>s_2 = 2 min.</td><td></td></tr>
<tr><td>n_1 = 20</td><td></td><td>n_2 = 20</td><td></td></tr>
</table>

Answer:
$H_0 : \mu_1 - \mu_2 = 0$; $H_1 : \mu_1 - \mu_2 \neq 0$
C.V. = ± 1.96
$t = 3.72$
Reject H_0. The time it takes to relieve a tension headache is different for the two brands.

Type: S   Sect: 3 Obj: 2   RANDOM: Y

44. A researcher wanted to know if the average number of phone calls received per day was the same in the city as in the country. Her hypothesis was there would be no difference in the number of calls. Test her claim at $\alpha = 0.01$. Assume variances are equal.

<table>
<tr><td>Country</td><td>City</td></tr>
<tr><td>$\bar{x}_1$ = 20</td><td>$\bar{x}_2$ = 13</td></tr>
<tr><td>s_1 = 3</td><td>s_2 = 2</td></tr>
<tr><td>n_1 = 20</td><td>n_2 = 25</td></tr>
</table>

Answer:
$H_0 : \mu_1 - \mu_2 = 0$; $H_1 : \mu_1 - \mu_2 \neq 0$
C.V. = ± 2.575
$t = 9.36$
Reject H_0. The average number of phone calls received in the city is different from the number of called received in the country.

Type: S   Sect: 3 Obj: 2   RANDOM: Y

45. A study was conducted to compare the speed limits in two different areas of town to see if there was any difference in the two. Use $\alpha = 0.05$ to test this claim. Assume variances are equal.

<u>Area 1</u>   <u>Area 2</u>

$\bar{x}_1 = 35$ mph $\bar{x}_2 = 25$ mph
$s_1 = 3$ mph $s_2 = 2$ mph
$n_1 = 16$ $n_2 = 18$

Answer:
$H_0 : \mu_1 - \mu_2 = 0$; $H_1 : \mu_1 - \mu_2 \neq 0$
C.V. $= \pm 1.96$
$t = 11.56$
Reject H_0. The speed limits are different.

Type: S   Sect: 3 Obj: 2   RANDOM: Y

46. A farmer feels that using a new fertilizer will increase the number of bushels per acre that he will produce. The table shows the number of bushels before and after applying the fertilizer. At $\alpha = 0.05$, did applying the fertilizer increase his yield?

Before	50	45	55	58	60	57	53
After	60	43	62	61	60	63	52

Answer:
$H_0 : \mu_D \geq 0$; $H_1 : \mu_D < 0$
C.V. $= -1.943$, d.f. $= 6$
$t = -1.92$
Do not reject H_0. There is not enough evidence to support the claim that the new fertilizer increases the farmer's yield.

Type: S   Sect: 4 Obj: 4   RANDOM: Y

47. A doctor felt that his patients' heart rates would decrease if they exercised. He selected 10 patients and recorded their heart rates, then placed them on an exercise plan for a month. The results are listed below. Test the doctor's hypothesis at $\alpha = 0.025$.

Before	80	86	75	74	72	70	85	74	72	68
After	75	80	75	69	70	72	79	72	68	68

Answer:
$H_0 : \mu_D \le 0$; $H_1 : \mu_D > 0$
C.V. = 2.262, d.f. = 9
$t = 3.14$
Reject H_0. It appears that the doctor's patients do have lower heart rates when they exercise.

Type: S   Sect: 4   Obj: 4   RANDOM: Y

48. A pollster believes that there will be no difference between voters' opinions of a candidate before a debate with his opponent and after the debate. Test this hypothesis at $\alpha = 0.05$. A higher numerical value represents a more favorable attitude toward the candidate.

Before	30	20	60	70	40	80	25	50	58
After	35	18	63	71	38	85	19	52	58

Answer:
$H_0 : \mu_D = 0$; $H_1 : \mu_D \neq 0$
C.V. = ±2.306, d.f. = 8
$t = -0.55$
Do not reject H_0. There appears to be no difference between voters' opinions of a candidate before and after a debate.

Type: S   Sect: 4   Obj: 4   RANDOM: Y

49. A researcher wanted to determine if there was a difference between the number of words typed per minute before a keyboarding course and after the course. She gathered the following data. Test her claim at $\alpha = 0.05$.

Before	35	37	40	55	60	32	75	80	43
After	50	40	45	55	70	50	73	81	60

Answer:
$H_0 : \mu_D = 0$; $H_1 : \mu_D \neq 0$
C.V. $= \pm 2.306$, d.f. $= 8$
$t = -2.89$
Reject H_0. There seems to be a difference between the number of words typed before and after a keyboarding course.

Type: S   Sect: 4 Obj: 4   RANDOM: Y

50. A consumer thinks the prices of dresses will decrease after Christmas. She kept track of the prices of 8 dresses both before and after Christmas. The results are listed below. Test her claim at $\alpha = 0.025$.

Before	85	95	110	55	60	75	65	90
After	75	65	95	55	60	50	45	75

Answer:
$H_0 : \mu_D \leq 0$; $H_1 : \mu_D > 0$
C.V. $= 2.365$, d.f. $= 7$
$t = 3.75$
Reject H_0. It appears that the consumer was right and the prices of dresses did decrease after Christmas.

Type: S   Sect: 4 Obj: 4   RANDOM: Y

51. A highway patrolman thinks that the speeds of motorists will decrease if he parks his car where they can see him. He first records the speeds of cars from an unseen vantage point and then parks his car in the open and records the speeds. The results are listed below. Test his claim at $\alpha = 0.025$.

Before	65	60	70	55	60	75	65	75	58
After	60	60	62	55	58	65	60	63	58

Answer:
$H_0 : \mu_D \leq 0$; $H_1 : \mu_D > 0$
C.V. $= 2.306$, d.f. $= 8$
Reject H_0. The highway patrolman's claim that motorists will slow down when he is in plain view seems to be correct.

Type: S   Sect: 4 Obj: 4   RANDOM: Y

52. Janice purchases a microwave oven. She hypothesizes that the amount of time it takes her to make dinner will be less than when she did not have a microwave. She records the amount of time it took her (in minutes) to prepare dinner both before and after purchasing the microwave. The data are listed in the table below. Test her hypothesis at $\alpha = 0.05$.

Before	55	45	60	48	50	60	65
After	40	40	55	35	48	60	58

Answer:
$H_0 : \mu_D \leq 0;\ H_1 : \mu_D > 0$
C.V. = 1.943, d.f. = 6
$t = 3.23$
Reject H_0. Janice seems to be correct in her assumption that using a microwave will decrease her preparation time for dinner.

Type: S   Sect: 4 Obj: 4   RANDOM: Y

CHAPTER 11

1. A scatter diagram is a histogram of the data.

   Answer: F Type: T   Sect: 1 Obj: 1   RANDOM: Y

2. A correlation coefficient of 0.98 would mean that as the values of x increase, the values of y increase also.

   Answer: T Type: T   Sect: 2 Obj: 2   RANDOM: Y

3. A pair of dice is rolled 6 times. The number of spots on the face of each die are listed below.

Die 1	5 4 6 2 3 6
Die 2	3 2 5 1 5 6

   The value of the correlation coefficient would be 0.35.

   Answer: F Type: T   Sect: 2 Obj: 2   RANDOM: Y

4. A correlation coefficient for 8 pieces of data was calculated to be 0.67. Using $\alpha = 0.05$, we can conclude that there is a significant relationship between the variables. (Assume a two-tailed test for significance.)

   Answer: F Type: T   Sect: 3 Obj: 3   RANDOM: Y

5. It was found that there was a significant relationship between two variables. We can conclude then that x causes y.

   Answer: F Type: T   Sect: 3 Obj: 3   RANDOM: Y

6. A regression line was calculated at $y' = 25.1 - 1.3x$. The slope of this line is -1.3.

   Answer: T Type: T   Sect: 4 Obj: 4   RANDOM: Y

7. A regression line was calculated as $y' = 30.1 - 0.52x$. If $x = 7$, the predicted value for y would be 34.2.

   Answer: F Type: T   Sect: 4 Obj: 5   RANDOM: Y

8. A multiple regression line was calculated as follows: $y' = 120.5 - 5.2x_1 + 3.1x_2$. If x_1 is 10 and x_2 is 8, the predicted value of $\hat{y}$ would be 93.3.

Answer: T Type: T   Sect: 13 Obj: 6   RANDOM: Y

9. A pair of dice is rolled 6 times. The number of spots on the face of each die are listed below.

Die 1	5	4	6	2	3	6
Die 2	3	2	5	1	5	6

The slope of the regression line is 0.5.

Answer: F Type: T   Sect: 4 Obj: 4   RANDOM: Y

10. A 95% confidence interval of the predicted value of y_f was calculated to be from 51.5 to 69.5. The predicted y_f value was 60.

Answer: F Type: T   Sect: 10 Obj: 7   RANDOM: Y

MULTIPLE CHOICE

11. The following data were collected:

x	6	7	10	5	7	8	12	5
y	8	10	8	9	9	12	15	10

The scatter diagram of these data would indicate what type of linear relationship?
a. negative
b. no linear relationship
c. positive
d. constant

Answer: c Type: M   Sect: 1 Obj: 1   RANDOM: Y

12. Which of the following could not be the value of a correlation coefficient?
a. 0.98
b. -0.76
c. -1.00
d. 1.34

Answer: d Type: M   Sect: 2 Obj: 2   RANDOM: Y

13. A correlation coefficient was computed for 10 sample points. The value of r was 0.68. Using the formula for the t-test, what is the test statistic that will be used to test the significance of the correlation coefficient?
    a. 2.78
    b. 2.62
    c. 3.21
    d. 10.12

Answer: b Type: M   Sect: 3 Obj: 3   RANDOM: Y

14. In order to find the equation of the line of best fit, the value of a was calculated as 20.5 and the value of b was calculated as -0.25. What is the equation of the line of best fit?
    a. $y' = -0.25 + 20.5x$
    b. $y' = 20.5 + 0.25x$
    c. $y' = -0.25 - 20.5x$
    d. $y' = 20.5 - 0.25x$

Answer: d Type: M   Sect: 4 Obj: 4   RANDOM: Y

15. What is the regression line for the following data points?

x	1	3	6	7	9	2
y	14	12	9	8	2	11

    a. $y' = 15.135 - 1.243x$
    b. $y' = 129.41 - 1.243x$
    c. $y' = 15.135 + 1.243x$
    d. $y' = -1.243 + 15.135x$

Answer: a Type: M   Sect: 6 Obj: 4   RANDOM: Y

16. A regression line was calculated as $y' = 42.5 + 3.1x$. What is the predicted value of y' when $x = 2$?
    a. 104.5
    b. 36.3
    c. 91.2
    d. 48.7

Answer: d Type: M   Sect: 6 Obj: 5   RANDOM: Y

17. If the correlation coefficient is 0.79, then what type of slope will the regression line have?
    a. positive
    b. negative
    c. constant
    d. zero

    Answer: a Type: M   Sect: 4 Obj: 4   RANDOM: Y

18. A multiple regression line was calculated as $y' = 35.7 + 2.5x_1 - 3.2x_2$. If $x_1 = 5$ and $x_2 = 10$, what is the predicted value of y'?
    a. 80.2
    b. 16.2
    c. 44.7
    d. 76.7

    Answer: b Type: M   Sect: 13 Obj: 6   RANDOM: Y

19. The following data list ages of cars and monthly repair cost.

Ages in Years	1	2	3	5	8	10
Monthly Cost	20	30	50	40	50	75

    The slope of the regression line for these data will have what type of slope?
    a. positive
    b. negative
    c. constant
    d. zero

    Answer: a Type: M   Sect: 4 Obj: 4   RANDOM: Y

20. What does R^2 measure?
    a. the amount of variation left to explain
    b. the proportion of variation in the dependent variable explained by the regression model
    c. how closely the errors fit a normal distribution
    d. the validity of the assumptions

    Answer: b Type: M   Sect: 9 Obj: 7   RANDOM: Y

SHORT ANSWER

Draw a scatter diagram for the given data.

21. The city of Oakdale wishes to see if there is a relationship between the temperature (in degrees Fahrenheit) and the amount of electricity used (in kilowatts). The data for the sample were:

Temperature (x)	68	73	82	95	90	85	80	100
Kilowatts (y)	800	750	900	1,500	1,200	1,000	900	1,600

Answer:

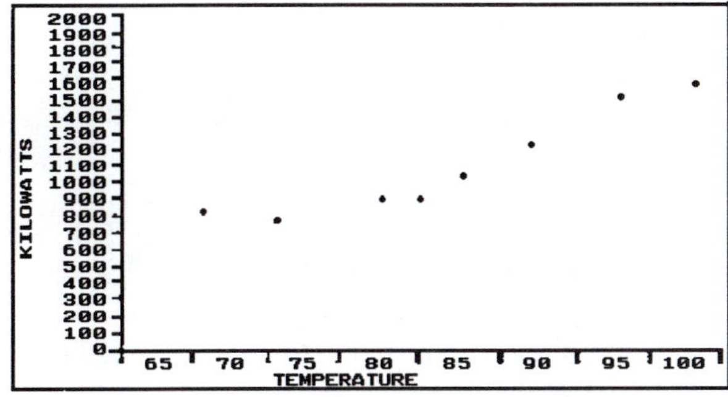

Type: S   Sect: 1 Obj: 1   RANDOM: Y

22. The following data show the demand for a product (in hundreds) and its price (in dollars) charged in 9 different cities.

Price (x)	25	15	18	20	19	23	21	17	22
Demand (y)	18	50	42	40	35	25	37	54	27

Answer:

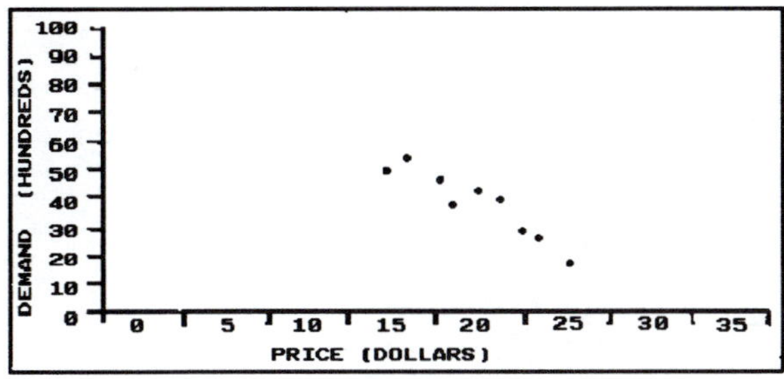

Type: S   Sect: 1 Obj: 1   RANDOM: Y

23. A researcher wanted to see if there is a relationship between a person's income (thousands of dollars) and the percent of that income given to charity. The following data were collected:

Income (*x*)	30	25	50	45	60	25	40	65	38	27
Percent (*y*)	5	4	3	3	4	3	3	5	3	3

Answer:

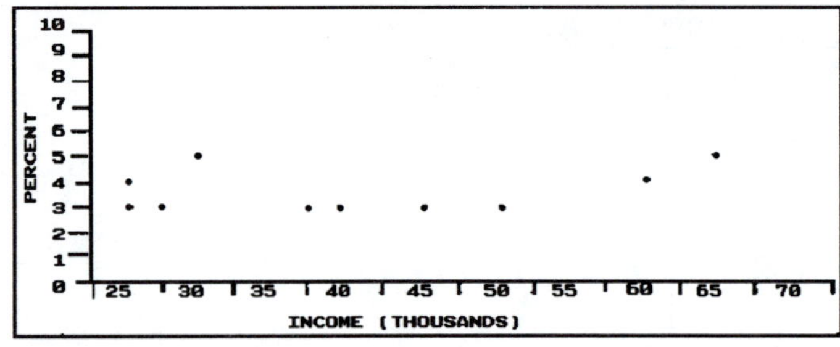

Type: S   Sect: 1 Obj: 1   RANDOM: Y

24. An obstetrician wants to know if there is a relationship between the age of his pregnant patients and the weight of their newborn babies. He collected the following data:

Mother's Age (*x*)	20	18	30	32	35
Baby's Weight (*y*)	7.5	7.2	6.5	6.2	6.3

Mother's Age (*x*)	24	28	31	19	26
Baby's Weight (*y*)	7.1	8.0	6.9	7.3	6.9

Answer:

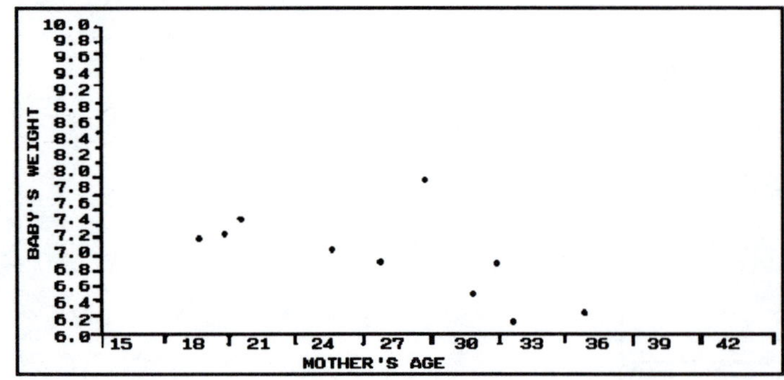

Type: S   Sect: 1 Obj: 1   RANDOM: Y

25. A science instructor wants to know if there is a relationship between the row a student sits in and the student's final grade in the course. A random sample of 12 students gave the following results.

Row (x)	1	2	4	3	2	1	4	3	3	2	4	1
Grade (y)	90	70	60	80	75	85	55	90	65	75	80	95

Answer:

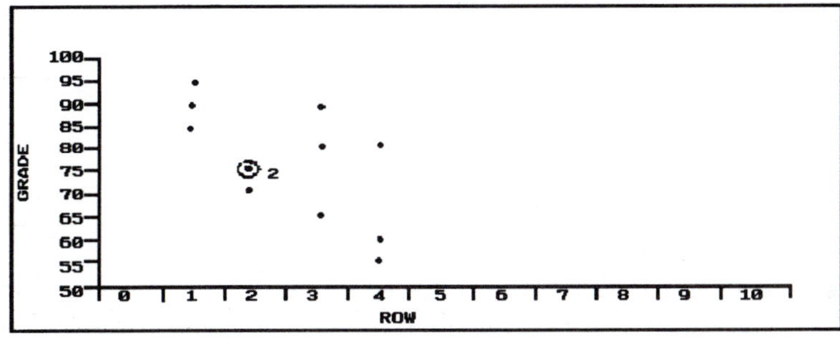

Type: S   Sect: 1 Obj: 1   RANDOM: Y

26. A study was conducted to determine if there was a linear relationship between a person's age and his or her peak heart rate. Data were collected from 8 people and are listed below.

Age (x)	15	25	30	35	40	50	45	20
Peak Heart Rate (y)	220	190	180	170	170	160	170	200

Answer:

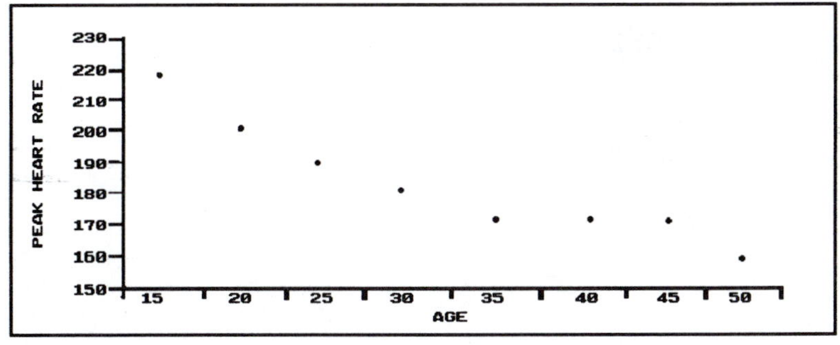

Type: S   Sect: 1 Obj: 1   RANDOM: Y

27. A high school weightlifting coach wanted to determine if there was a relationship between the weight of his wrestlers and the number of pounds they could bench press. The data are listed below.

Weight (x)	120	145	160	180	200	135	190	150
Bench Press (y)	220	230	270	280	320	210	300	240

Answer:

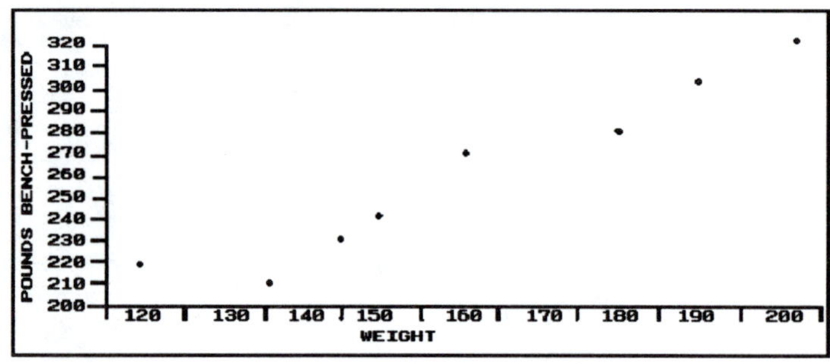

Type: S   Sect: 1 Obj: 1   RANDOM: Y

Compute the value of the correlation coefficient for the given data.

28. The city of Oakdale wishes to see if there is a relationship between the temperature (in degrees Fahrenheit) and the amount of electricity used (in kilowatts). The data for the sample were:

Temperature (x)	68	73	82	95	90	85	80	100
Kilowatts (y)	800	750	900	1,500	1,200	1,000	900	1,600

Answer: $r = 0.943$ Type: S   Sect: 2 Obj: 2   RANDOM: Y

29. The following data show the demand for a product (in hundreds) and its price (in dollars) charged in 9 different cities.

Price (x)	25	15	18	20	19	23	21	17	22
Demand (y)	18	50	42	40	35	25	37	54	27

Answer: $r = -0.94$ Type: S   Sect: 2 Obj: 2   RANDOM: Y

30. A researcher wanted to see if there is a relationship between a person's income (thousands of dollars) and the percent of that income given to charity. The following data were collected:

Income (x)	30	20	50	45	60	25	40	65	38	27
Percent (y)	5	4	3	3	4	3	3	5	3	3

Answer: $r = 0.262$ Type: S   Sect: 2 Obj: 2   RANDOM: Y

31. An obstetrician wants to know if there is a relationship between the age of his pregnant patients and the weight of their newborn babies. He collected the following data:

Mother's Age (x)	20	18	30	32	35
Baby's Weight (y)	7.5	7.2	6.5	6.2	6.3

Mother's Age (x)	24	28	31	19	26
Baby's Weight (y)	7.1	8.0	6.9	7.3	6.9

Answer: $r = -0.624$ Type: S   Sect: 2 Obj: 2   RANDOM: Y

32. A science instructor wants to know if there is a relationship between the row a student sits in and the student's final grade in the course. A random sample of 12 students gave the following results:

Row (x)	1	2	4	3	2	1	4	3	3	2	4	1
Grade (y)	90	70	60	80	75	85	55	90	65	75	80	95

Answer: $r = -0.654$ Type: S   Sect: 2 Obj: 2   RANDOM: Y

33. A study was conducted to determine if there was a linear relationship between a person's age and his or her peak heart rate. Data were collected from 8 people and are listed below.

Age (x)	15	25	30	35	40	50	45	20
Peak Heart Rate (y)	220	190	180	170	170	160	170	200

Answer: $r = -0.94$ Type: S   Sect: 2 Obj: 2   RANDOM: Y

34. A high school weightlifting coach wanted to determine if there was a relationship between the weight of his wrestlers and the number of pounds they could bench press. The data are listed below.

Weight (x)	120	145	160	180	200	135	190	150
Bench Press (y)	200	230	270	280	320	210	300	240

Answer: $r = 0.987$ Type: S   Sect: 2 Obj: 2   RANDOM: Y

Test the significance of the correlation coefficient at $\alpha = 0.05$ for the given data and give a brief explanation of the type of relationship. (Use a two-tailed hypothesis test).

35. The city of Oakdale wishes to see if there is a relationship between the temperature (in degrees Fahrenheit) and the amount of electricity used (in kilowatts). The data for the sample were ($r = 0.943$):

Temperature (x)	68	73	82	95	90	85	80	100
Kilowatts (y)	800	750	900	1,500	1,200	1,000	900	1,600

Answer:
$H_0 : \wp = 0;\ H_1 : \wp \neq 0$
$r = 0.943$
test statistic: 6.94; C.V. $= \pm 2.447$; d.f. $= 6$
Reject H_0. There is probably a relationship between the temperature and the amount of electricity used.

Type: S   Sect: 3 Obj: 3   RANDOM: Y

36. A science instructor wants to know if there is a relationship between the row a student sits in and the student's final grade in the course. A random sample of 12 students gave the following results ($r = -0.654$).

Row (x)	1	2	4	3	2	1	4	3	3	2	4	1
Grade (y)	90	70	60	80	75	85	55	90	65	75	80	95

Answer:
$H_0 : \wp = 0;\ H_1 : \wp \neq 0$
$r = -0.654$
test statistic: -2.99; C.V. $= \pm 2.228$; d.f. $= 10$
Reject H_0. There is probably a relationship between the row a student sits in and the student's final grade.

Type: S   Sect: 3 Obj: 3   RANDOM: Y

37. A study was conducted to determine if there was a linear relationship
between a person's age and his or her peak heart rate. Data were
collected from 8 people and are listed below ($r = -0.94$).

Age (x)	15	25	30	35	40	50	45	20
Peak Heart Rate (y)	220	190	180	170	170	160	170	200

Answer:
$H_0 : \rho = 0$; $H_1 : \rho \neq 0$
$r = -0.94$
test statistic: -6.75; C.V. = ±2.447; d.f. = 6
Reject H_0. There is probably a relationship between a patient's age and peak heart
rate.

Type: S   Sect: 3 Obj: 3   RANDOM: Y

38. A high school weightlifting coach wanted to determine if there was a
relationship between the weight of his wrestlers and the number of
pounds they could bench press. The data are listed below
($r = 0.987$).

Weight (x)	120	145	160	180	200	135	190	150
Bench Press (y)	200	230	270	280	320	210	300	240

Answer:
$H_0 : \rho = 0$; $H_1 : \rho \neq 0$
$r = -0.987$
test statistic: 15.04; C.V. = ±2.447; d.f. = 6
Reject H_0. There is probably a relationship between the weight of wrestlers and how
much weight they can bench press.

Type: S   Sect: 3 Obj: 3   RANDOM: Y

39. What would you conclude if 40 pairs of data produce a linear correlation coefficient of
$r = -0.72$, if $\alpha = 0.05$?

Answer: It appears that there is a significant negative relationship between the two variables.

Type: S   Sect: 2 Obj: 3   RANDOM: Y

40. What would you conclude if 18 pairs of data produce a linear correlation coefficient of
$r = 0.82$, if $\alpha = 0.05$?

Answer: There is probably a positive relationship between the two variables.   Type: S

Sect: 2 Obj: 3   RANDOM: Y

41. What would you conclude if 12 pairs of data produce a linear correlation coefficient of $r = -0.67$, if $\alpha = 0.01$?

    Answer: There is not a relationship between the two variables. Type: S   Sect: 2

    Obj: 3   RANDOM: Y

For the following data, compute the value of the correlation coefficient, test the significance of the correlation coefficient at $\alpha = 0.05$, and give a brief explanation of the type of relationship. (Use a 2-tailed test).

42. A consumer advocate decided to find out if there was a linear relationship between the age of a car and its annual repair costs. She conducted a survey of 9 car owners and obtained the following results:

Age of Car (x)	2	4	6	5	10	5	4	3	6
Repair Costs (y)	40	75	180	175	570	150	125	80	400

    Answer:
    $H_0 : \rho = 0; \; H_1 : \rho \neq 0$
    $r = 0.926$
    test statistic: 6.49; C.V. $= \pm 2.365$; d.f. $= 7$
    Reject H_0. There is probably a relationship between the age of a car and its annual repair cost.

    Type: S   Sect: 3 Obj: 2   RANDOM: Y

43. A neighborhood wanted to know if the number of policemen patrolling the streets would be linearly related to the number of robberies. The following data were obtained:

No. of Policemen (x)	5	8	10	7	9	4	9	12	10	8
No. of Robberies (y)	30	25	18	25	23	35	20	10	15	22

    Answer:
    $H_0 : \rho = 0; \; H_1 : \rho \neq 0$
    $r = 0.973$
    test statistic: -11.92; C.V. $= \pm 2.306$; d.f. $= 8$
    Reject H_0. There appears to be a significant relationship between the number of policemen patrolling the streets and the number of robberies.

    Type: S   Sect: 3 Obj: 2   RANDOM: Y

44. A banker thinks there is a linear relationship between the interest rates on home loans and the number of homes his bank finances. He keeps records for the next few months and obtains the following results:

Interest Rate—% (x)	10	10.5	11	12	9.5	11.5
Homes Financed (y)	65	65	50	35	90	45

Answer:
$H_0 : \rho = 0$; $H_1 : \rho \neq 0$
$r = -0.964$
test statistic: -7.25; C.V. = ±2.776; d.f. = 4
Reject H_0. There seems to be a significant negative relationship between the interest rates on home loans and the number of homes financed.

Type: S   Sect: 3 Obj: 2   RANDOM: Y

45. A psychologist wants to determine if there is a linear relationship between the number of hours a person goes without sleep and the number of mistakes he or she makes on a simple test. She records the following data:

Hrs. Without Sleep (x)	30	36	48	24	42	30	36	42
No. of Mistakes (y)	6	8	15	7	10	7	10	12

Answer:
$H_0 : \rho = 0$; $H_1 : \rho \neq 0$
$r = 0.903$
test statistic: 5.15; C.V. = ±2.447; d.f. = 6
Reject H_0. There appears to be a significant relationship between the number of hours a person goes without sleep and the number of mistakes made on a simple test.

Type: S   Sect: 3 Obj: 2   RANDOM: Y

46. A sociologist wants to see if there is a linear relationship between a person's age and his or her attitude toward older people. A scale of from 1 to 50 is used to measure attitudes toward older people, with 50 being the worst attitude. The results are listed below.

Age (x)	18	25	40	35	52	61	48	28	55
Attitude (y)	45	40	28	30	10	5	14	20	15

Answer:
$H_0 : \rho = 0$; $H_1 : \rho \neq 0$
$r = -0.905$
test statistic: -5.63; C.V. = ±2.365; d.f. = 7
Reject H_0. There appears to be a negative relationship between a person's age and his or her attitude toward older people.

Type: S   Sect: 3 Obj: 2   RANDOM: Y

47. A study is done to determine if there is a linear relationship between a person's weight at birth and at age 25. The results are listed below.

Birth Weight (x)	7.5	6.5	6.4	8.5	9.2	7.2	6.1
Weight at 25 (y)	145	130	125	169	155	135	120

Answer:
$H_0 : \rho = 0$; $H_1 : \rho \neq 0$
$r = 0.91$
test statistic: 4.91; C.V. = ±2.571; d.f. = 5
Reject H_0. There appears to be a positive relationship between a person's weight at birth and at age 25.

Type: S   Sect: 3 Obj: 2   RANDOM: Y

Find the equation of the regression line for the following data.

48. The city of Oakdale wishes to see if there is a relationship between the temperature (in degrees Fahrenheit) and the amount of electricity used (in kilowatts). The data for the sample were as follows:

Temperature (x)	68	73	82	95	90	85	80	100
Kilowatts (y)	800	750	900	1,500	1,200	1,000	900	1,600

Answer: $y' = -1280.916 + 28.08x$ Type: S   Sect: 6 Obj: 4   RANDOM: Y

49. The following data show the demand for a product (in hundreds) and its price (in dollars) charged in 9 different cities.

Price (x)	25	15	18	20	19	23	21	17	22
Demand (y)	18	50	42	40	35	25	37	54	27

Answer: $y' = 106.957 - 3.526x$ Type: S   Sect: 6 Obj: 4   RANDOM: Y

50. A consumer advocate decided to find out if there was a linear relationship between the age of a car and its annual repair costs. She conducted a survey of 9 car owners and obtained the following results:

Age of Car (x)	2	4	6	5	10	5	4	3	6
Repair Costs (y)	40	75	180	175	570	150	125	80	400

Answer: $y' = -151.746 - 70.238x$ Type: S   Sect: 6 Obj: 4   RANDOM: Y

51. A neighborhood wanted to know if the number of policemen patrolling the streets would be linearly related to the number of robberies. The following data were obtained:

No. of Policemen (x)	5	8	10	7	9	4	9	12	10	8
No. of Robberies (y)	30	25	18	25	23	35	20	10	15	22

Answer: $y' = 46.23 - 2.92x$ Type: S   Sect: 6 Obj: 4   RANDOM: Y

52. A banker thinks there is a linear relationship between the interest rates on home loans and the number of homes his bank finances. He keeps records for the next few months and obtains the following results:

Interest Rate—% (x)	10	10.5	11	12	9.5	11.5
Homes Financed (y)	65	65	50	35	90	45

Answer: $y' = 273.33 - 20x$ Type: S   Sect: 6 Obj: 4   RANDOM: Y

53. A psychologist wants to determine if there is a linear relationship between the number of hours a person goes without sleep and the number of mistakes he or she makes on a simple test. She records the following data:

Hrs. Without Sleep (x)	30	36	48	24	42	30	36	42
No. of Mistakes (y)	6	8	15	7	10	7	10	12

Answer: $y' = -3.125 + 0.347x$ Type: S   Sect: 6 Obj: 4   RANDOM: Y

54. A sociologist wants to see if there is a linear relationship between a person's age and his or her attitude toward older people. A scale of from 1 to 50 is used to measure attitudes toward older people, with 50 being the worst attitude. The results are listed below.

Age (x)	18	25	40	35	52	61	48	28	55
Attitude (y)	45	40	28	30	10	5	14	20	15

Answer: $y' = 56.619 - 0.836x$ Type: S   Sect: 6 Obj: 4   RANDOM: Y

Use the regression line to predict a given value.

55. If $y' = 1.56 + 3.5x$, what is the predicted value of y' when $x = 12$?

Answer: 43.56 Type: S   Sect: 6 Obj: 5   RANDOM: Y

56. If $y' = 30.5 - 1.89x$, what is the predicted value of y' when $x = 4$?

Answer: 22.94 Type: S   Sect: 6 Obj: 5   RANDOM: Y

57. If $y' = -3.5 + 4.6x$, what is the predicted value of y' when $x = 25$?

Answer: 111.5 Type: S   Sect: 6 Obj: 5   RANDOM: Y

58. If $y' = 120.6 - 3.4x$, what is the predicted value of y' when $x = 30$?

Answer: 18.6 Type: S   Sect: 6 Obj: 5   RANDOM: Y

59. A banker thinks there is a linear relationship between the interest rates on home loans and the number of homes his bank finances. He keeps records for the next few months and obtains the following results:

Interest Rate—% (x)	10	10.5	11	12	9.5	11.5
Homes Financed (y)	65	65	50	35	90	45

What is the predicted number of homes financed when the interest rate is 11.75%?

Answer: approximately 38 homes Type: S   Sect: 6 Obj: 5   RANDOM: Y

60. A psychologist wants to determine if there is a linear relationship between the number of hours a person goes without sleep and the number of mistakes he or she makes on a simple test. She records the following data:

Hrs. Without Sleep (x)	30	36	48	24	42	30	36	42
No. of Mistakes (y)	6	8	15	7	10	7	10	12

What is the predicted number of mistakes made by a person who goes without 32 hours of sleep?

Answer: 8 mistakes Type: S   Sect: 6   Obj: 5   RANDOM: Y

61. A study was conducted to determine if there was a linear relationship between a person's age and his or her peak heart rate. Data were collected from 8 people and are listed below.

Age (x)	15	25	30	35	40	50	45	20
Peak Heart Rate (y)	220	190	180	170	170	160	170	200

What is the predicted peak heart rate for a person at age 33?

Answer: approximately 182 Type: S   Sect: 6 Obj: 5   RANDOM: Y

62. A high school weightlifting coach wanted to determine if there was a relationship between the weight of his wrestlers and the number of pounds they could bench press. The data are listed below.

Weight (x)	120	145	160	180	200	135	190	150
Bench Press (y)	200	230	270	280	320	210	300	240

What is the predicted number of pounds bench pressed by a wrestler who weighs 176 pounds?

Answer: approximately 280 pounds Type: S   Sect: 6 Obj: 5   RANDOM: Y

63. The city of Oakdale wishes to see if there is a relationship between the temperature (in degrees Fahrenheit) and the amount of electricity used (in kilowatts). The data for the sample were as follows:

Temperature (x)	68	73	82	95	90	85	80	100
Kilowatts (y)	800	750	900	1,500	1,200	1,000	900	1,600

What is the predicted number of kilowatts used when the temperature is 98°?

Answer: approximately 1,471 kilowatts Type: S   Sect: 6 Obj: 5   RANDOM: Y

64. The following data show the demand for a product (in hundreds) and its price (in dollars) charged in 9 different cities.

Price (x)	25	15	18	20	19	23	21	17	22
Demand (y)	18	50	42	40	35	25	37	54	27

What is the demand for a product if the price is $23?

Answer: approximately 2,586 products or, to the nearest hundred, 2,600 products.

Type: S   Sect: 6 Obj: 5   RANDOM: Y

Use the given multiple regression line to predict values.

65. A multiple regression line was calculated as $y' = 35.7 + 2.5x_1 - 3.2x_2$. If $x_1 = 6$ and $x_2 = 12$, what is the predicted value of y'?

Answer: 12.3 Type: S   Sect: 13 Obj: 6   RANDOM: Y

66. A multiple regression line was calculated as $y' = 120.5 - 5.2x_1 + 3.1x_2$. If x_1 is 15 and x_2 is 18, what is the predicted value of y'?

   Answer: 98.3 Type: S   Sect: 13 Obj: 6   RANDOM: Y

67. A multiple regression line was calculated as follows: $y' = 250 - 8.3x_1 + 4.5x_2$. If x_1 is 20 and x_2 is 25, what is the predicted value of y'?

   Answer: 196.5 Type: S   Sect: 13 Obj: 6   RANDOM: Y

68. A multiple regression line was calculated in which x_1 is the annual family income (in thousands of dollars) and x^2 is the number of people in the family. The multiple regression line was calculated as $y' = 250 - 6.5x_1 + 4.5x_2$. If the annual family income is \$36,000 and there are 3 people in the family, what is the predicted value of y'?

   Answer: 29.5 Type: S   Sect: 13 Obj: 6   RANDOM: Y

69. A multiple regression line was calculated in which x_1 is the annual family income (in thousands of dollars) and x^2 is the number of people in the family. The multiple regression line was calculated as: $y' = 250 - 4.6x_1 + 4.5x_2$. If the annual family income is \$50,000 and there are 4 people in the family, what is the predicted value of y'?

   Answer: 38 Type: S   Sect: 13 Obj: 6   RANDOM: Y

70. A multiple regression line was calculated in which x_1 is a student's grade-point average and x_2 is a student's age. The multiple regression line was calculated as: $y' = -42.5 + 90.5x_1 + 13.65x_2$. If a student has a grade-point average of 3.5 and is 25 years old, what is the predicted value of y'?

   Answer: 615.5 Type: S   Sect: 13 Obj: 6   RANDOM: Y

71. A multiple regression line was calculated in which x_1 is a student's grade-point average and x_2 is a student's age. The multiple regression line was calculated as: $y' = -42.5 + 90.5x_1 + 13.65x_2$. If a student has a grade-point average of 3.2 and is 30 years old, what is the predicted value of y'?

   Answer: 656.6 Type: S   Sect: 13 Obj: 6   RANDOM: Y

72. In testing the correlation between two variables, $r = 0.76$ and the test statistic was 3.70. How many items were in the sample?

   Answer: 12 Type: S   Sect: 3 Obj: 3   RANDOM: Y

73. In testing the correlation between two variables, $r = 0.65$ and the test statistic was 3.63. How many items were in the sample?

    Answer: 20 Type: S   Sect: 3 Obj: 3   RANDOM: Y

74. What is a reason for performing regression analysis?

    Answer:
    to determine if variables are related, to measure the relationship between the variables, or to predict a value.
    Type: S   Sect: 4 Obj: 4   RANDOM: Y

75. Consider the equation $N = 42.7 + 6.5A$, where N = number of items sold and A = advertisements placed. Which variable is the dependent variable?

    Answer: N is the dependent variable. Type: S   Sect: 4 Obj: 4   RANDOM: Y

76. If you believe that X has a negative effect on Y, which of the following is appropriate?
    a. $H_1 : b > 0$ b. $H_1 : b < 0$
    c. $H_1 : b \neq 0$ d. $H_0 : b = 0$

    Answer: b Type: S   Sect: 9 Obj: 4   RANDOM: Y

77. The estimated regression line is $y' = 3.46 + 0.88x$, where y = sales and x = advertising.

Advertising (x)	Sales (y)
2	1
4	0
6	1
8	5

    Find a 95% prediction interval for y_f if $x_f = 10$ and $\hat{\sigma}_e = 2.07$.

    Answer: -1.82 to 26.34 Type: S   Sect: 10 Obj: 7   RANDOM: Y

78. The estimated regression line is $y' = 3.46 + 0.88x$, where $y =$ sales and $x =$ advertising.

Advertising (x)	Sales (y)
2	1
4	0
6	1
8	5

Find a 90% prediction interval for y_f if $x_f = 5$ and $\hat{\sigma}_e = 1.32$.

Answer: 3.55 to 12.17 Type: S   Sect: 10 Obj: 7   RANDOM: Y

79. The estimated regression line is $y' = 273.33 - 20x$ and the standard error of estimate is 5.77, where the following data were used:

x	10	10.5	11	12	9.5	11.5
y	65	65	50	35	90	45

Find a 95% prediction interval for y_f if $x_f = 11$.

Answer: 35.92 to 70.74 Type: S   Sect: 10 Obj: 7   RANDOM: Y

80. The estimated regression line is $y' = 273.33 - 20x$ and the standard error of estimate is 5.77, where the following data were used:

x	10	10.5	11	12	9.5	11.5
y	65	65	50	35	90	45

Find a 90% prediction interval for y_f if $x_f = 11.5$.

Answer: 29.33 to 57.33 Type: S   Sect: 10 Obj: 7   RANDOM: Y

81. The estimated regression line is $y' = 273.33 - 20x$ and the standard error of estimate is 5.65, $\sum(x - \bar{x})^2 = 4.5$, $\bar{x} = 10.75$, and $n = 7$. Find a 98% prediction interval for y_f when $x_f = 10.25$.

Answer: 47.52 to 89.14 Type: S   Sect: 10 Obj: 7   RANDOM: Y

82. The estimated regression line is $y' = -3.125 + 0.347x$ and the standard error of estimate is 1.40, where the following data were used:

x	30	36	48	24	42	30	36	42
y	6	8	15	7	10	7	10	12

Find a 95% prediction interval for y_f if $x_f = 40$.

Answer: 7.06 to 14.45 Type: S   Sect: 10 Obj: 7   RANDOM: Y

83. The estimated regression line is $y' = -3.125 + 0.347x$ and the standard error of estimate is 1.40, where the following data were used:

x	30	36	48	24	42	30	36	42
y	6	8	15	7	10	7	10	12

Find a 98% prediction interval for y_f if $x_f = 36$.

Answer: 4.7 to 14.03 Type: S   Sect: 10 Obj: 7   RANDOM: Y

84. The estimated regression line is $y' = -3.125 + 0.347x$, and the standard error of estimate is 1.40, $\Sigma(x - \bar{x})^2 = 432$, $\bar{x} = 36$, and $n = 8$. Find a 90% prediction interval for y_f when $x_f = 44$.

Answer: 9.07 to 15.21 Type: S   Sect: 10 Obj: 7   RANDOM: Y

85. The estimated regression line is $y' = 232.024 - 1.524x$ and the standard error of estimate is 7.21, where the following data were used:

x	15	25	30	35	40	50	45	20
y	220	190	180	170	170	160	170	200

Find a 98% prediction interval for y_f if $x_f = 45$.

Answer: 137.9 to 189 Type: S   Sect: 10 Obj: 7   RANDOM: Y

86. The estimated regression line is $y' = 232.024 - 1.524x$ and the standard error of estimate is 7.21, where the following data were used:

x	15	25	30	35	40	50	45	20
y	220	190	180	170	170	160	170	200

Find a 95% prediction interval for y_f if $x_f = 30$.

Answer: 167.5 to 205.06 Type: S   Sect: 10 Obj: 7   RANDOM: Y

87. The estimated regression line is $y' = 232.024 - 1.524x$, the standard error of estimate is 7.21, $\Sigma(x - \bar{x})^2 = 1050$, $\bar{x} = 32.5$, and $n = 8$. Find a 99% prediction interval for y_f when $x = 35$.

Answer: 150.26 to 207.11 Type: S   Sect: 10 Obj: 7   RANDOM: Y

88. Using $y = 8.22 + 0.376x$, $\Sigma(x - \bar{x})^2 = 750$, the standard error of estimate = 2.5, $\bar{x} = 30$, and $n = 10$, what is the 95% prediction interval for y_f when $x = 32$?

Answer: 14.19 to 26.31 Type: S   Sect: 10 Obj: 7   RANDOM: Y

****
The remaining questions in this chapter require the use of the formula generation capabilities of ESATEST III. Please see the ESATEST III documentation if you need directions on how to use them.
****

89. If $y' = @1.56 + @3.5x$, what is the predicted value of y' when $x = @12$?

Answer: @1+@2*@3,1.56,1,2,2,3.5,3,5,0,12,10,20,0 Type: S   RANDOM: F

MULTIPLE CHOICE

90. In order to find the equation of the line of best fit, the value of a was calculated as @20.5 and the value of b was calculated as ` @-0.25`. What is the equation of the line of best fit?
a. $y' = {\sim}2 + {\sim}1x$
b. $y' = {\sim}1 - {\sim}2x$
c. $y' = {\sim}2 - {\sim}1x$
d. $y' = {\sim}1 + {\sim}2x$

Answer: d,20.5,20,30,1,-0.25,-0.3,-0.2,2 Type: M   RANDOM: F

91. A regression line was calculated as $y' = @42.5 + @3.1x$. What is the predicted value of y' when $x = @2$?
    a. |104.5
    b. |36.3
    c. |91.2
    d. |48.7

Answer: @1+@2*@3,42.5,40,45,1,3.1,3,5,1,2,2,9,0   Type: M   RANDOM: F

92. A multiple regression line was calculated as $y' = @35.7 + 2.5x_1 - @3.2x_2$. If $x_1 = 5$ and $x_2 = @10$, what is the predicted value of y'?
    a. |80.2
    b. |16.2
    c. |44.7
    d. |76.7

Answer: 12.5+@1-@2*@3,35.7,30,40,1,3.2,3,4,1,10,5,10,0   Type: M   RANDOM: F

CHAPTER 12

TRUE/FALSE

1. To analyze the effect of several different values of a variable or of several different sets of circumstances of a controlled experiment, we can use the technique analysis of variance.

   Answer: T Type: T   Sect: 1 Obj: 2   RANDOM: Y

2. The value of the F-distribution can be negative.

   Answer: F Type: T   Sect: 3 Obj: 1   RANDOM: Y

3. A factor is a variable that can potentially change the outcome.

   Answer: T Type: T   Sect: 1 Obj: 2   RANDOM: Y

4. Experimental design is the process of selecting participants, assigning them to treatment samples, and obtaining prescribed information from them.

   Answer: T Type: T   Sect: 1 Obj: 2   RANDOM: Y

5. The null hypothesis of ANOVA is that the means of all populations are not equal.

   Answer: F Type: T   Sect: 1 Obj: 2   RANDOM: Y

6. The initials SSB stand for "sum of squares within."

   Answer: F Type: T   Sect: 1 Obj: 2   RANDOM: Y

7. If H_0 is true and all of the population means are equal, the sum-of-squares value, SSW, should be smaller.

   Answer: F Type: T   Sect: 1 Obj: 2   RANDOM: Y

8. A random variable with an F probability distribution is discrete.

   Answer: F Type: T   Sect: 3 Obj: 1   RANDOM: Y

9. Suppose a random variable, F, has 9 degrees of freedom in the numerator and 12 degrees of freedom in the denominator. Using $\alpha = 0.025$, the critical value is 3.8%.

   Answer: F Type: T   Sect: 3 Obj: 1   RANDOM: Y

10. A random variable, F, has 12 degrees of freedom in the numerator and 10 degrees of freedom in the denominator. Using $\alpha = 0.01$, the critical value is 4.71.

    Answer: T Type: T   Sect: 3 Obj: 1   RANDOM: Y

11. If the F table does not include a specific degrees-of-freedom value required in a problem, then use the closest value that is smaller than the desired value.

    Answer: T Type: T   Sect: 3 Obj: 1   RANDOM: Y

12. If an F statistic has 20 degrees of freedom in the numerator and 9 degrees of freedom in the denominator, the proportion of F values that exceed 2.94 is 0.025.

    Answer: F Type: T   Sect: 3 Obj: 1   RANDOM: Y

13. If an F statistic has 28 degrees of freedom in the numerator and 20 degrees of freedom in the denominator, the proportion of F values that exceed 2.86 is 0.01.

    Answer: T Type: T   Sect: 3 Obj: 1   RANDOM: Y

14. If the F-ratio is 3.76, based on 30 observations from 4 different samples, and $\alpha = 0.025$, we would reject H_0.

    Answer: T Type: T   Sect: 4 Obj: 2   RANDOM: Y

15. If the F-ratio is 2.87, based on 28 observations from 3 different samples, and $\alpha = 0.05$, we would reject H_0.

    Answer: F Type: T   Sect: 4 Obj: 2   RANDOM: Y

16. If the F-ratio is 3.07, based on 32 observations from 5 different samples, and $\alpha = 0.025$, we would reject H_0.

    Answer: F Type: T   Sect: 4 Obj: 2   RANDOM: Y

17. If the F-ratio is 4.67, based on 35 observations from 4 different samples, and $\alpha = 0.01$, we would reject H_0.

    Answer: T Type: T   Sect: 4 Obj: 2   RANDOM: Y

18. If the F-ratio is 3.97, based on 31 observations from 4 different samples, and $\alpha = 0.025$, we would reject H_0.

    Answer: T Type: T   Sect: 4 Obj: 2   RANDOM: Y

19. The $P(F_{(10,15)} > 2.54) = 0.05$.

Answer: T Type: T   Sect: 3 Obj: 1   RANDOM: Y

20. The $P(F_{(20,25)} > 2.70) = 0.025$.

Answer: F Type: T   Sect: 3 Obj: 1   RANDOM: Y

MULTIPLE CHOICE

21. A store manager wants to see if there is a difference in the average time (in minutes) a customer has to wait in line for 4 of his checkers. The times are shown below.

Checker 1	Checker 2	Checker 3	Checker 4
3	3	2	2
5	5	3	3
4	2	5	4
6	3	6	4
7	4	3	5

How many degrees of freedom in the denominator?
a. 20   b. 16   c. 15   d. 3

Answer: b Type: M   Sect: 4 Obj: 2   RANDOM: Y

22. A store manager wants to see if there is a difference in the average time (in minutes) a customer has to wait in line for 4 of his checkers. The times are shown below.

Checker 1	Checker 2	Checker 3	Checker 4
3	3	2	2
5	5	3	3
4	2	5	4
6	3	6	4
7	4	3	5

What is the grand mean?
a. 5   b. 3.4   c. 3.8   d. 3.95

Answer: d Type: M   Sect: 1 Obj: 2   RANDOM: Y

23. What is the critical value for a one-tailed right F-test when $\alpha = 0.01$, the degrees of freedom for the numerator = 12, and the degrees of freedom for the denominator = 18?
a. 1.93
b. 3.86
c. 3.37
d. 3.46

Answer: c Type: M   Sect: 3 Obj: 1   RANDOM: Y

24. What is the critical value for an F-test with $\alpha = 0.025$, the degrees of freedom for the numerator = 15, and the degrees of freedom for the denominator = 18?
    a. 2.86   b. 2.67   c. 2.72   d. 2.62

    Answer: b Type: M Sect: 3 Obj: 1   RANDOM: Y

25. A history teacher has 3 sections of the same course. He wants to determine if there is a difference between the averages of the three sections. The grades on 5 exams are listed below.

Section A	Section B	Section C
72	68	69
75	74	75
70	72	73
78	70	73
79	73	75

    What is the grand mean for these data?
    a. 73.0667
    b. 72.6
    c. 72.9576
    d. 73.7215

    Answer: a Type: M Sect: 1 Obj: 2   RANDOM: Y

26. Which of the following terms is not used when calculating the test statistic for analysis of variance?
    a. grand mean   b. chi-square value
    c. sum of squares within   d. mean sum of squares within

    Answer: b Type: M Sect: 4 Obj: 2   RANDOM: Y

27. What is the value of $P(F_{(20, 15)} > 2.76)$?
    a. 0.05   b. 0.025   c. 0.01   d. 0.005

    Answer: b Type: M Sect: 3 Obj: 1   RANDOM: Y

28. What is the value of $P(F_{(13, 25)} > 3.37)$?
    a. 0.05   b. 0.025   c. 0.01   d. 0.005

    Answer: d Type: M Sect: 3 Obj: 1   RANDOM: Y

29. What is the value of $P(F_{(9, 12)} > 4.39)$?
    a. 0.05   b. 0.025   c. 0.01   d. 0.005

    Answer: c Type: M Sect: 3 Obj: 1   RANDOM: Y

30. What is the critical value for a one-tailed right F-test when $\alpha = 0.025$, the degrees of freedom for the numerator = 13, and the degrees of freedom for the denominator = 18?

Answer: 2.77 Type: M   Sect: 3 Obj: 1   RANDOM: Y

31. What is the critical value for a one-tailed right F-test when $\alpha = 0.01$, the degrees of freedom for the numerator = 10, and the degrees of freedom for the denominator = 20?

Answer: 3.37 Type: M   Sect: 3 Obj: 1   RANDOM: Y

32. What is the critical value for a one-tailed right F-test when $\alpha = 0.05$, the degrees of freedom for the numerator = 20, and the degrees of freedom for the denominator = 15?

Answer: 2.33   Type: M   Sect: 3 Obj: 1   RANDOM: Y

33. What is the critical value for a one-tailed right F-test when $\alpha = 0.005$, the degrees of freedom for the numerator = 15, and the degrees of freedom for the denominator = 13?

Answer: 4.46 Type: M   Sect: 3 Obj: 1   RANDOM: Y

34. What is the critical value for a one-tailed right F-test when $\alpha = 0.025$, the degrees of freedom for the numerator = 10, and the degrees of freedom for the denominator = 25?

Answer: 2.61 Type: M   Sect: 3 Obj: 1   RANDOM: Y

35. What is the value of f such that $P(F_{(15, 12)} > f) = 0.01$?
 a. 4.25   b. 4.01   c. 3.82   d. 3.67

Answer: b Type: M   Sect: 3 Obj: 1   RANDOM: Y

36. What is the value of f such that $P(F_{(20, 15)} > f) = 0.05$?
 a. 3.88   b. 2.33   c. 2.20   d. 2.38

Answer: b Type: M   Sect: 3 Obj: 1   RANDOM: Y

37. What is the value of f such that $P(F_{(23, 20)} > f) = 0.025$?
 a. 2.36   b. 2.41   c. 2.46   d. 2.51

Answer: c Type: M   Sect: 3 Obj: 1   RANDOM: Y

38. What is the value of f such that $P(F_{(20, 28)} > f) = 0.01$?
 a. 2.60   b. 2.63   c. 2.57   d. 1.69

Answer: a Type: M   Sect: 3 Obj: 1   RANDOM: Y

39. What is the value of f such that $P(F_{(8, 32)} > f) = 0.005$?
    a. 2.27   b. 6.40   c. 3.35   d. 3.58

    Answer: d Type: M Sect: 3 Obj: 1 RANDOM: Y

40. What is the value of f such that $P(F_{(16, 30)} > f) = 0.01$?
    a. 2.55   b. 2.70   c. 3.49   d. 3.75

    Answer: b Type: M Sect: 3 Obj: 1 RANDOM: Y

41. The ANOVA procedure will be used to test the difference between 4 means. The number of pieces of data for each of the samples is as follows: group A--7, group B--8, group C--10, and group D--12. What is the number of degrees of freedom of the denominator?
    a. 3   b. 37   c. 35   d. 33

    Answer: d Type: M Sect: 4 Obj: 2 RANDOM: Y

42. The ANOVA procedure will be used to test the difference between 5 means. The number of pieces of data for each of the samples is as follows: group A--6, group B--8, group C--10, group D--8, and group E--7. What is the number of degrees of freedom of the denominator?
    a. 34   b. 39   c. 36   d. 35

    Answer: a Type: M Sect: 4 Obj: 2 RANDOM: Y

43. The ANOVA procedure will be used to test the difference between 3 means. The number of degrees of freedom of the denominator is 32. The number of items in group 1 is 9, and the number of items in group 2 is 12. How many items are in group 3?
    a. 15   b. 20   c. 18   d. 14

    Answer: d Type: M Sect: 4 Obj: 2 RANDOM: Y

44. A researcher wants to compare 3 different means to see if there is any difference between them. Group A contains 9 items, group B 10 items, and group C 9 items. If $\alpha = 0.01$, what is the critical value?
    a. 5.53   b. 5.49   c. 4.64   d. 5.42

    Answer: a Type: M Sect: 4 Obj: 2 RANDOM: Y

SHORT ANSWER

45. What is $P(F_{(30, 20)} > 2.78)$?

    Answer: 0.01 Type: S Sect: 3 Obj: 1 RANDOM: Y

46.What is $P(F_{(10, 15)} > 3.06)$?

    Answer: 0.025 Type: S   Sect: 3 Obj: 1   RANDOM: Y

47.What is $P(F_{(18, 20)} > 3.50)$?

    Answer: 0.005 Type: S   Sect: 3 Obj: 1   RANDOM: Y

48.What is $P(F_{(12, 8)} > 3.28)$?

    Answer: 0.05 Type: S   Sect: 3 Obj: 1   RANDOM: Y

49.What is $P(F_{(9, 12)} > 3.44)$?

    Answer: 0.025 Type: S   Sect: 3 Obj: 1   RANDOM: Y

50.What is $P(F_{(20, 25)} > 2.70)$?

    Answer: 0.01 Type: S   Sect: 3 Obj: 1   RANDOM: Y

51.What is $P(F_{(5, 12)} > 5.06)$?

    Answer: 0.01 Type: S   Sect: 3 Obj: 1   RANDOM: Y

52.What is $P(F_{(50, 30)} > 2.52)$?

    Answer: 0.005 Type: S   Sect: 3 Obj: 1   RANDOM: Y

53.What is $P(F_{(22, 16)} > 4.99)$?

    Answer: 0.001 Type: S   Sect: 3 Obj: 1   RANDOM: Y

54.What is the value of f such that $P(F_{(15, 20)} > f) = 0.05$?

    Answer: 2.20 Type: S   Sect: 3 Obj: 1   RANDOM: Y

55.What is the value of f such that $P(F_{(23, 18)} > f) = 0.005$?

    Answer: 3.50 Type: S   Sect: 3 Obj: 1   RANDOM: Y

56.What is the value of f such that $P(F_{(3, 26)} > f) = 0.01$?

    Answer: 4.64 Type: S   Sect: 3 Obj: 1   RANDOM: Y

57. What is the value of f such that $P(F_{(20, 15)} > f) = 0.05$?

    Answer: 2.33 Type: S   Sect: 3 Obj: 1   RANDOM: Y

58. What is the value of f such that $P(F_{(4, 28)} > f) = 0.025$?

    Answer: 3.29 Type: S   Sect: 3 Obj: 1   RANDOM: Y

59. What is the value of f such that $P(F_{(11, 17)} > f) = 0.01$?

    Answer: 3.59 Type: S   Sect: 3 Obj: 1   RANDOM: Y

60. What is the value of f such that $P(F_{(27, 30)} > f) = 0.0$

    Answer: 2.47 Type: S   Sect: 3 Obj: 1   RANDOM: Y

61. What is the value of f such that $P(F_{(30, 20)} > f) = 0.05$?

    Answer: 2.04 Type: S   Sect: 3 Obj: 1   RANDOM: Y

62. What is the value of f such that $P(F_{(13, 25)} > f) = 0.025$?

    Answer: 2.51 Type: S   Sect: 3 Obj: 1   RANDOM: Y

63. When using the F-test to compare four different groups, the total number of items in the four groups was 28. What is the degrees of freedom for the denominator associated with these data?

    Answer: 24 Type: S   Sect: 4 Obj: 2   RANDOM: Y

64. When using the F-test to compare four different groups, the total number of items in the four groups was 24. If $\alpha = 0.01$, what is the critical value for these data?

    Answer: 4.94 Type: S   Sect: 4 Obj: 2   RANDOM: Y

65. When using the F-test to compare three different groups, the total number of items in the three groups was 22. If $\alpha = 0.025$, what is the critical value for these data?

    Answer: 4.51 Type: S   Sect: 4 Obj: 2   RANDOM: Y

66. When using the F-test to compare five different groups, the total number of items in the five groups was 28. If $\alpha = 0.05$, what is the critical value for these data?

    Answer: 2.80 Type: S   Sect: 4 Obj: 2   RANDOM: Y

67. When using the F-test to compare three different groups, the total number of items in the three groups was 20. If $\alpha = 0.01$, what is the critical value for these data?

    Answer: 6.11 Type: S   Sect: 4 Obj: 2   RANDOM: Y

68. When using the F-test to compare five different groups, the total number of items in the five groups was 37. If $\alpha = 0.025$, what is the critical value for these data?

    Answer: 3.25 Type: S   Sect: 4 Obj: 2   RANDOM: Y

69. When using the F-test to compare four different groups, the total number of items in the four groups was 30. What is the degrees of freedom for the denominator associated with these data?

    Answer: 26 Type: S   Sect: 4 Obj: 2   RANDOM: Y

70. When using the F-test to compare five different groups, the total number of items in the five groups was 29. What is the degrees of freedom for the denominator associated with these data?

    Answer: 24 Type: S   Sect: 4 Obj: 2   RANDOM: Y

71. A carpenter needs to purchase some new tools. He wants to determine if there is any difference in price between 3 different hardware stores in the city. He decides to randomly pick out 7 tools and check their price at each of the 3 stores. The prices (in dollars) are listed below.

Cooper's	Bob's	Big John's
40	42	39
35	34	38
15	14	15
12	13	14
30	28	31
50	55	54
25	23	23

What is the grand mean?

Answer: 30 Type: S   Sect: 1 Obj: 2   RANDOM: Y

72. A botanist plants random samples of each of three different strains of roses on 6 plots of ground. The plots are the same size and fertility. The yields per plot are as follows:

<u>Strain 1</u>	<u>Strain 2</u>	<u>Strain 3</u>
12	9	12
10	15	15
9	14	16
11	12	20
10	13	18
13	8	11

What is the grand mean?

Answer: 12.6667 Type: S   Sect: 1 Obj: 2   RANDOM: Y

73. A Christmas tree retailer wants to determine if there is a difference in number of sales between the types of Christmas tree selected. He recorded the number of sales for 5 days and the results were as follows:

<u>Cedar</u>	<u>Pine</u>	<u>Spruce</u>
10	12	14
8	10	12
13	8	10
15	14	15
9	13	12

What is the grand mean?

Answer: 11.6667 Type: S   Sect: 1 Obj: 2   RANDOM: Y

74. A customer wants to decide if there is a difference in the average price of a long-distance call between US Sprint, AT&T, and MCI. She compares prices of calls at various times of day using each of three companies. The results were as follows:

<u>US Sprint</u>	<u>AT&T</u>	<u>MCI</u>
2.50	2.60	2.45
1.25	1.15	1.30
5.10	5.20	5.05
10.00	9.80	9.75
4.00	4.15	4.05
6.50	6.70	6.65

What is the grand mean?

Answer: 4.9 Type: S   Sect: 1 Obj: 2   RANDOM: Y

75. A Christmas tree retailer wants to determine if there is a difference in number of sales between the types of Christmas tree selected. He recorded the number of sales for 5 days and the results were as follows:

Cedar	Pine	Spruce
10	12	14
8	10	12
13	8	10
15	14	15
9	13	12

What is the sum of squares between, SSB?

Answer: 6.9217   Type: S   Sect: 4 Obj: 2   RANDOM: Y

76. If we want to see if there is any difference in price between 3 different types of Christmas trees, and the sum of squares between $(SSB) = 6.9217$, what is the mean sum of squares between equal?

Answer: 3.4609 Type: S   Sect: 4 Obj: 2   RANDOM: Y

77. If we want to see if there is any difference in price between 3 different types of Christmas trees, where $SST = 79.3217$ and SSB = 6.9217, what is the sum of squares within (SSW)?

Answer: 72.4 Type: S   Sect: 4 Obj: 2   RANDOM: Y

78. If we want to see if there is any difference in price between 3 different types of Christmas trees, where $MSB = 3.46$ and $SSW = 72.4$, what is the test statistic $(n = 15)$?

Answer: 0.5736 Type: S   Sect: 4 Obj: 2   RANDOM: Y

79. If we want to see if there is any difference in price between 3 different types of Christmas trees, where $n = 15$ and the test statistic is 0.78, what decision would we make $(\alpha = 0.025)$?

Answer: We cannot reject H_0, since $0.78 < 5.10$. Type: S   Sect: 4 Obj: 2

RANDOM: Y

80. If we want to see if there is any difference between 4 different means, where $MSB = 16.205$, $n = 25$, and $SSW = 100.2$, what is the test statistic?

Answer: 3.3963 Type: S   Sect: 4 Obj: 2   RANDOM: Y

81. If we want to determine if there is any difference between 4 different means, where $n = 27$, $\alpha = 0.01$, and the test statistic is 3.5692, what decision should we make?

    Answer: We cannot reject H_0, since $3.5692 < 4.76$.   Type: S   Sect: 4   Obj: 2

    RANDOM: Y

82. If we want to determine if there is any difference between 5 different means, where $n = 32$, the test statistic is 4.627, and $\alpha = 0.025$, what decision should we make?

    Answer: We reject H_0, since $4.627 > 3.31$.   Type: S   Sect: 4   Obj: 2   RANDOM: Y

83. If we want to determine if there is any difference between 5 different means, where $n = 35$, the test statistic is 3.967, and $\alpha = 0.05$, what decision should we make?

    Answer: We reject H_0, since $3.967 > 2.69$ Type: S   Sect: 4   Obj: 2   RANDOM: Y

84. If you want to determine if there is any difference in grading between 4 of your teachers, where $\alpha = 0.025$, $n = 32$, and the test statistic is 3.5612, what decision would you make?

    Answer: You cannot reject H_0, since $3.5612 < 3.63$.   Type: S   Sect: 4   Obj: 2

    RANDOM: Y

85. We want to reject H_0 for a test between 4 different means, where $n = 25$, and the test statistic is 3.5963. What is the smallest value that α could equal in order to reject H_0?

    Answer: $\alpha = 0.05$ Type: S   Sect: 4   Obj: 2   RANDOM: Y

86. If we want to determine if there is any difference in effectiveness between four different kinds of medicine, and the sum of the squares between $(SSB) = 48.721$, what is the mean sum of squares between?

    Answer: 16.2403   Type: S   Sect: 4   Obj: 2   RANDOM: Y

87. If we want to determine if there is any difference in effectiveness between four different kinds of medicine, where $n = 37$, the test statistic is 3.6721, and $\alpha = 0.05$, what decision should be made?

    Answer: We reject H_0, since $3.6721 > 2.92$.   Type: S   Sect: 4   Obj: 2   RANDOM: Y

88. If we want to determine if there is any difference in effectiveness between four different kinds of medicine, where $SST = 273.57$ and $SSB = 56.215$, what is the sum of squares within?

    Answer: 217.355 Type: S   Sect: 4 Obj: 2   RANDOM: Y

89. If we want to determine if there is any difference in effectiveness between four different kinds of medicine, where $n = 35$, $SSW = 237.526$, and $MSB = 20.176$, what is the test statistic?

    Answer: 2.6332 Type: S   Sect: 4 Obj: 2   RANDOM: Y

90. If we want to determine if there is any difference in effectiveness between four different kinds of medicine, where $n = 32$, $\alpha = 0.025$, and the test statistic is 2.723, what decision should be made?

    Answer: We cannot reject H_0, since $2.723 < 3.63$. Type: S   Sect: 4 Obj: 2

    RANDOM: Y

91. To determine if there is a difference between three means, where the grand mean was 11.25, $n = 25$, and $SS = 3{,}521.231$, what is the total sum of squares?

    Answer: 357.1685 Type: S   Sect: 4 Obj: 2   RANDOM: Y

92. To determine if there is a difference between three means, and $SSB = 261.6125$, what is the mean of squares between?

    Answer: 130.8063 Type: S   Sect: 4 Obj: 2   RANDOM: Y

93. To determine if there is a difference between three means, where $SSB = 256.1719$, $SST = 358.1219$, and $n = 22$, what is the mean sum of squares within?

    Answer: 5.3658 Type: S   Sect: 4 Obj: 2   RANDOM: Y

94. What is the test statistic if $MSB = 17.6258$ and $MSW = 5.2671$?

    Answer: 3.3464 Type: S   Sect: 4 Obj: 2   RANDOM: Y

95. To determine if there is a difference between three means, where $n = 23$ and the test statistic is 3.4216, what decision should be made ($\alpha = 0.025$)?

    Answer: We cannot reject H_0, since $3.4216 < 4.46$. Type: S   Sect: 4 Obj: 2

    RANDOM: Y

96.To determine if there is a difference between five means, where
$n = 40$ and the test statistic is 3.789, what decision should be made for $\alpha = 0.01$?

Answer: We cannot reject H_0, since $3.789 < 4.02$.   Type: S   Sect: 4   Obj: 2

RANDOM: Y

Use the following problem to answer the questions below.

A history teacher has 3 sections of the same course. He wants to determine if there is a
difference between the averages of the three sections.   The grades on 5 exams are listed
below.

Section A	Section B	Section C
72	68	69
75	74	75
70	72	73
78	70	73
79	73	75

97.What is the grand mean?

Answer: 73.0667 Type: S   Sect: 4   Obj: 2   RANDOM: Y

98.What is the sum of squares?

Answer: 80,216 Type: S   Sect: 4   Obj: 2   RANDOM: Y

99.If $SS = 80,216$, what is the total sum of squares?

Answer: 134.8603 Type: S   Sect: 4   Obj: 2   RANDOM: Y

100.If $SST = 134.8603$ and $SSB = 28.8603$, what is the mean sum of squares within?

Answer: 8.8333 Type: S   Sect: 4   Obj: 2   RANDOM: Y

101.What is the sum of squares between (SSB)?

Answer: 28.8603 Type: S   Sect: 4   Obj: 2   RANDOM: Y

102.If $SSB = 28.8603$ and $SSW = 106$, what is the test statistic?

Answer: 1.6336 Type: S   Sect: 4   Obj: 2   RANDOM: Y

103. If the test statistic is 1.6336 and $\alpha = 0.05$, what decision should be made?

   Answer: We cannot reject H_0, since $1.6336 < 3.81$. Type: S   Sect: 4 Obj: 2

   RANDOM: Y

Use the following problem to answer the questions below.

Burger Chef has outlets in four different sections of town. The district manager wants to determine if there is a difference in satisfaction level between the four outlets. Customers are given questionnaires and rate the outlets on a scale of 1 to 10, with 10 the highest rating. The results are listed below.

Outlet 1	Outlet 2	Outlet 3	Outlet 4
7	8	9	8
7	7	8	7
6	8	9	7
5	8	9	6
6	8	8	6
6	7	9	7

104. What is the grand mean?

   Answer: 7.3333 Type: S   Sect: 4 Obj: 2   RANDOM: Y

105. What is the total sum of squares?

   Answer: 29.3451 Type: S   Sect: 4 Obj: 2   RANDOM: Y

106. If $SST = 29.3451$ and $SSB = 21.018$, what is the mean sum of squares between?

   Answer: 7.006 Type: S   Sect: 4 Obj: 2   RANDOM: Y

107. What is the sum of squares between (SSB)?

   Answer: 21.0179 Type: S   Sect: 4 Obj: 2   RANDOM: Y

108. What is the critical value for this problem ($\alpha = 0.01$)?

   Answer: 4.94 Type: S   Sect: 4 Obj: 2   RANDOM: Y

109. If $SSB = 21.0179$ and $SSW = 8.3272$, what is the test statistic?

   Answer: 16.8251 Type: S   Sect: 4 Obj: 2   RANDOM: Y

110. If the test statistic is 16.83 and $\alpha = 0.01$, what decision should be made?

Answer: We reject H_0, since $16.83 > 4.94$. Type: S   Sect: 4 Obj: 2   RANDOM: Y

Use the folloiwng problem to answer the questions below.

A researcher wants to determine if there is a difference in the amount of time a person has to wait in line at three banks in town. He records the waiting taime (in minutes) of randomly selected people at the three banks. The results are as follows:

National Bank	State Bank	Commerce Bank
5	4.5	3.5
6	3.5	4
5.5	3	3
5	4.5	3.5
5. 5	4.5	4
	4	4

111. What is the grand mean?

Answer: 4.2941 Type: S   Sect: 4 Obj: 2   RANDOM: Y

112. What is the total sum of squares?

Answer: 12.532 Type: S   Sect: 4 Obj: 2   RANDOM: Y

113. What is the sum of squares between (SSB)?

Answer: 9.0001 Type: S   Sect: 4 Obj: 2   RANDOM: Y

114. If $SSB = 9.0001$, what is the mean sum of squares between?

Answer: 4.5001 Type: S   Sect: 4 Obj: 2   RANDOM: Y

115. What is the critical value for this problem if $\alpha = 0.025$?

Answer: 4.86 Type: S   Sect: 4 Obj: 2   RANDOM: Y

116. If $MSB = 4.5$ and $SSW = 3.532$, what is the test statistic?

Answer: 17.8359 Type: S   Sect: 4 Obj: 2   RANDOM: Y

117. If the test statistic is 17.84 and $\alpha = 0.025$, what decision should be made?

Answer: We reject H_0, since $17.84 > 4.86$. Type: S   Sect: 4 Obj: 2   RANDOM: Y

Use the following problem to answer the questions below.

Three different brands of cigarettes were tested for tar content to determine if there is a difference between the brands. The following table gives the tar content (in milligrams) of the three brands.

Brand 1	Brand 2	Brand 3
344	330	322
352	325	328
360	335	335
333	315	317
328	318	317

118. What is the grand mean?

Answer: 330.6 Type: S   Sect: 4 Obj: 2   RANDOM: Y

119. What is the total sum of squares?

Answer: 2,437.6 Type: S   Sect: 4 Obj: 2   RANDOM: Y

120. What is the sum of squares between (SSB)?

Answer: 1,230.4 Type: S   Sect: 4 Obj: 2   RANDOM: Y

121. If $SST = 2437.6$ and $SSB = 1230.4$, what is the mean sum of squares within?

Answer: 100.6 Type: S   Sect: 4 Obj: 2   RANDOM: Y

122. If $MSB = 615.2$, what is SSB?

Answer: 1,230.4 Type: S   Sect: 4 Obj: 2   RANDOM: Y

123. If $MSB = 615.2$ and $SSW = 1,207.2$, what is the test statistic?

Answer: 6.1153 Type: S   Sect: 4 Obj: 2   RANDOM: Y

124. What is the critical value if $\alpha = 0.01$?

Answer: 6.93 Type: S Sect: 4 Obj: 2   RANDOM: Y

125. What decision should be made if $\alpha = 0.01$ and the test statistic is 6.12?

Answer: We cannot reject H_0, since $6.12 < 6.93$. Type: S   Sect: 4 Obj: 2

RANDOM: Y

Use the following problem to answer the questions below.

A diet center wanted to test 3 different methods of losing weight to determine if the average weight loss (in pounds) per week for each method was the same. The results of the methods are as follows:

<u>Liquid</u> <u>Diet</u>	<u>Low-Calorie</u> <u>Diet</u>	<u>Low-Calorie</u> <u>Diet</u> <u>&</u> <u>Exercise</u>
5	3	3
2	2	3
1	3	4
2	2	2.5
2.5	1.5	3
4	2	3
1.5	1.5	2

126.What is the sum of squares (SS)?

   Answer: 155.25 Type: S   Sect: 4 Obj: 2   RANDOM: Y

127.If SS = 155.25, what is the total sum of squares?

   Answer: 18.9544 Type: S   Sect: 4 Obj: 2   RANDOM: Y

128.What is the sum of squares between (SSB)?

   Answer: 2.1701 Type: S   Sect: 4 Obj: 2   RANDOM: Y

129.If SST = 18.9544 and SSB = 2.1701, what is the mean sum of squares within?

   Answer: 0.9325 Type: S   Sect: 4 Obj: 2   RANDOM: Y

130.What is the critical value if α = 0.05?

   Answer: 3.55 Type: S   Sect: 4 Obj: 2   RANDOM: Y

131.If MSB = 1.0851 and SSW = 16.7843, what is the test statistic?

   Answer: 1.1636 Type: S   Sect: 4 Obj: 2   RANDOM: Y

132.What decision should be made if α = 0.05 and the test statistic is 1.164?

   Answer: We cannot reject H_0, since 1.164 < 3.55. Type: S   Sect: 4 Obj: 2

   RANDOM: Y

Use the following problem to answer the questions below.

A botanist plants random samples of each of three different strains of roses on 6 plots of ground. The plots are the same size and fertility. The yields per plot are as follows:

Strain 1	Strain 2	Strain 3
12	9	12
10	15	15
9	14	16
11	12	20
10	13	18
13	8	11

133. What is the total sum of squares?

Answer: 175.9848 Type: S   Sect: 4 Obj: 2   RANDOM: Y

134. What is the sum of squares between (SSB)?

Answer: 66.9695 Type: S   Sect: 4 Obj: 2   RANDOM: Y

135. If $SST = 175.985$ and $SSB = 66.97$, what is the mean sum of squares within?

Answer: 7.2677 Type: S   Sect: 4 Obj: 2   RANDOM: Y

136. If $MSB = 33.4848$ and $SSW = 109.0153$, what is the test statistic?

Answer: 4.6073 Type: S   Sect: 4 Obj: 2   RANDOM: Y

137. What decision should be made if $\alpha = 0.025$ and the test statistic is 4.6073?

Answer: We cannot reject H_0, since $4.6073 < 4.77$. Type: S   Sect: 4 Obj: 2

RANDOM: Y

138. If we test statistic is 4.61 and $\alpha = 0.05$, what decision should be made?

Answer: We reject H_0, since $4.61 > 3.68$. Type: S   Sect: 4 Obj: 2   RANDOM: Y

139. A botanist plants random samples of each of three different strains of roses on 6 plots of ground. The plots are the same size and fertility. The yields per plot are as follows:

Strain 1	Strain 2	Strain 3
12	9	12
10	15	15
9	14	16
11	12	20
10	13	18
13	8	11

If we reject H_0 at $\alpha = 0.05$, what conclusion could we make?

Answer:
It would appear that there is a difference in yield of roses, depending on the strain selected.

Type: S   Sect: 4   Obj: 2   RANDOM: Y

140. A customer wants to decide if there is a difference in the average price of a long-distance phone call between US Sprint, AT&T and MCI. She compares prices of calls at various times of day using each of the 3 companies. At $\alpha = 0.05$, she does not reject the null hypothesis. What conclusion did she make?

Answer:
There did not appear to be any significant difference between the 3 companies in the average price of a long-distance telephone call.

Type: S   Sect: 4   Obj: 2   RANDOM: Y

141. Four different brands of cigarettes were tested for tar content to determine if there was a difference between the brands. If the test statistic was 4.65 and the critical value was 3.89, what decision was made?

Answer: We reject H_0, since 4.65 > 3.89.   Type: S   Sect: 4   Obj: 2   RANDOM: Y

142. If $n = 12$, $J = 3$, and $\alpha = 0.01$, what is the critical value?

Answer: 8.02 Type: S   Sect: 4   Obj: 2   RANDOM: Y

MULTIPLE CHOICE

```
****************************************************************
```
The remaining questions in this chapter require the use of the formula generation capabilities of ESATEST III. Please see the ESATEST III documentation if you need directions on how to use them.
```
****************************************************************
```

143. The ANOVA procedure will be used to test the difference between 4 means. The number of pieces of data for each of the samples is as follows: group A—@7, group B—@8, group C—@10, and group D—@12` . What is the number of degrees of freedom of the denominator?

    a. I 3   b. I 37   c. I 35   d. I 33

    Answer: @1+@2+@3+@4-4,7,1,9,0,8,1,9,0,10,10,20,0,12,10,20,0 Type: M   RANDOM: F

SHORT ANSWER

144. When using the F-test to compare four different groups, the total number of items in the four groups was `@28` . What is the degrees of freedom for the denominator associated with these data?

    Answer: @1-4,28,10,40,0 Type: S   RANDOM: F

145. If we want to see if there is any difference between @4 different means, where $MSB = $ @16.205, $n = $ @25, and $SSW = $ @100.2, what is the test statistic?

    Answer: @2*(@3-@1)/@4,4,2,6,0,16.205,10,20,3,25,20,40,0,100,100,200,1,[4]   Type: S

    RANDOM: F

146. To determine if there is a difference between three means, where the grand mean was @11.25, $n = $ @25, and $SS = $ @3521.231, what is the total sum of squares?

    Answer: @3-@2*@1^2,11.25,10,11.5,2,25,10,25,0,3521.231,3500,4000,3,[4]   Type: S

    RANDOM: F

147. To determine if there is a difference between three means, and $SSB = $ @261.6125, what is the mean of squares between?

    Answer: @1/2,261.6125,200,300,4,[4]   Type: S   RANDOM: F

148. To determine if there is a difference between three means, where $SSB = @256.1719$, $SST = @358.1219$, and $n = @22$, what is the mean sum of squares within?

Answer: (@2-@1)/(@3-3),256,200,300,4,358,300,400,0,22,20,40,0,[4]   Type: S   RANDOM: F

149. What is the test statistic if $MSB = @17.6258$ and $MSW = @5.2671$?

Answer: @1/@2,17,10,20,4,5,1,9,4,[4]   Type: S   RANDOM: F

MULTIPLE CHOICE

1. The chi-square test of independence is a test that demonstrates a cause and effect between two variables.

   Answer: F Type: M   Sect: 2 Obj: 2   RANDOM: Y

2. The critical value for a chi-square distribution with 14 degrees of freedom when α = 0.01 and the test is one-tailed to the right is 26.1.

   Answer: F Type: M   Sect: 3 Obj: 1   RANDOM: Y

3. The critical chi-square value for 13 degrees of freedom when α = 0.01 and the test is one-tailed right is 26.2.

   Answer: F Type: M   Sect: 3 Obj: 1   RANDOM: Y

4. For a valid result, the chi-square test for independence requires that the expected frequencies be at least five for each cell.

   Answer: T Type: M   Sect: 2 Obj: 2   RANDOM: Y

5. If the differences between the observed and expected frequencies in a contingency table are generally small, we attribute them to sampling error.

   Answer: T Type: M   Sect: 3 Obj: 2   RANDOM: Y

6. A chi-square distribution is always positive.

   Answer: T Type: M   Sect: 3 Obj: 1   RANDOM: Y

7. $P(x^2{}_{(17)} > 27.6) = 0.05$

   Answer: T Type: M   Sect: 3 Obj: 1   RANDOM: Y

8. $P(x^2{}_{(8)} > 20.1) = 0.025$

   Answer: F Type: M   Sect: 3 Obj: 1   RANDOM: Y

9. The chi-square goodness-of-fit test is always one-tailed right.

   Answer: T Type: M   Sect: 3 Obj: 2   RANDOM: Y

10. A chi-square test of independence was performed to determine if the type of crime committed in a large city and the area where the crime was committed were independent of each other. The decision was to reject H_0. This means that the type of crime committed was independent of the area in which the crime took place.

Answer: F Type: M   Sect: 3 Obj: 2   RANDOM: Y

11. Consider the following set of data.

	Response	
	Yes	No
Group 1	30	10
Group 2	20	15
Group 3	40	20

The expected frequency for group 3 with a response of yes is 40.

Answer: T Type: M   Sect: 2 Obj: 2   RANDOM: Y

12. A contingency table has 4 rows and 3 columns. The degrees of freedom would be 12.

Answer: F Type: M   Sect: 3 Obj: 2   RANDOM: Y

13. What is the chi-square critical value for 13 degrees of freedom when $\alpha = 0.01$ and the test is one-tailed right?
    a. 24.7
    b. 22.4
    c. 27.7
    d. 19.8

Answer: c Type: M   Sect: 3 Obj: 1   RANDOM: Y

14. What is the value of c so that $P(x^2_{(10)} > c) = 0.025$?
    a. 18.3   b. 20.5   c. 23.2   d. 25.2

Answer: b Type: M   Sect: 3 Obj: 1   RANDOM: Y

15. What is $P(x^2_{(12)} > 23.3)$?
    a. 0.05   b. 0.025   c. 0.01   d. 0.005

Answer: b Type: M   Sect: 3 Obj: 1   RANDOM: Y

16. What is $P(x^2_{(18)} > 34.8)$?
    a. 0.05   b. 0.025   c. 0.01   d. 0.005

Answer: c Type: M   Sect: 3 Obj: 1   RANDOM: Y

17. What is $P(x^2_{(20)} > 31.4)$?
    a. 0.05   b. 0.025   c. 0.01   d. 0.005

    Answer: a Type: M   Sect: 3 Obj: 1   RANDOM: Y

18. What is $P(x^2_{(11)} > 17.3)$?
    a. 0.05   b. 0.025   c. 0.10   d. 0.01

    Answer: c Type: M   Sect: 3 Obj: 1   RANDOM: Y

19. What is the value of c so that $P(x^2_{(15)} > c) = 0.05$?
    a. 25.0   b. 27.5   c. 30.6   d. 32.8

    Answer: a Type: M   Sect: 3 Obj: 1   RANDOM: Y

20. What is the value of c so that $P(x^2_{(19)} > c) = 0.025$?
    a. 30.1   b. 36.2   c. 38.6   d. 32.9

    Answer: d Type: M   Sect: 3 Obj: 1   RANDOM: Y

21. What is the value of c so that $P(x^2_{(9)} > c) = 0.01$?
    a. 14.7   b. 16.9   c. 19.0   d. 21.7

    Answer: d Type: M   Sect: 3 Obj: 1   RANDOM: Y

22. A chi-square test of independence is to be performed with a contingency table that has 4
    rows and 5 columns. How many degrees of freedom are there?
    a. 20
    b. 12
    c. 16
    d. 15

    Answer: b Type: M   Sect: 3 Obj: 2   RANDOM: Y

23. A researcher selects a random sample of items from 3 different shopping centers. He rates the prices of the items as low, moderate, or high. The results are shown below.

Prices

	Low	Moderate	High
Center 1	18	20	16
Center 2	8	18	30
Center 3	14	25	12

What is the expected frequency of Center 3 with a moderate price on their items?
a. 19.9565
b. 12.6708
c. 22.6087
d. 21.1304

Answer: a Type: M Sect: 2 Obj: 2   RANDOM: Y

24. A researcher selects a random sample of items from 3 different shopping centers. He rates the prices of the items as low, moderate, or high. The results are shown below.

Prices

	Low	Moderate	High
Center 1	18	20	16
Center 2	8	18	30
Center 3	14	25	12

What is the expected frequency of Center 1 with a moderate price?
a. 19.9565
b. 22.6087
c. 21.1304
d. 20.6271

Answer: c Type: M Sect: 2 Obj: 2   RANDOM: Y

25.The favorite colors of males and females are listed below.

	Red	Blue	Green	Yellow
Male	20	30	30	10
Female	10	40	25	15

What is the expected frequency for a male who likes blue?
a. 30   b. 40   c. 35   d. 25

Answer: c Type: M   Sect: 2 Obj: 2   RANDOM: Y

26.The favorite colors of males and females are listed below.

	Red	Blue	Green	Yellow
Male	20	30	30	10
Female	10	40	25	15

What is the expected frequency for a female who prefers red?
a. 10   b. 15   c. 20   d. 25

Answer: b Type: M   Sect: 2 Obj: 2   RANDOM: Y

27.The favorite colors of males and females are listed below.

	Red	Blue	Green	Yellow
Male	20	30	30	10
Female	10	40	25	15

What is the expected frequency for a male who prefers red?
a. 10   b. 20   c. 15   d. 25

Answer: c Type: M   Sect: 2 Obj: 2   RANDOM: Y

28.The favorite colors of males and females are listed below.

	Red	Blue	Green	Yellow
Male	20	30	30	10
Female	10	40	25	15

What is the expected frequency for a male who prefers green?
a. 25   b. 30   c. 23.5   d. 27.5

Answer: d Type: M   Sect: 2 Obj: 2   RANDOM: Y

SHORT ANSWER

29.What is $P(x^2_{(17)} > 24.8)$?

   Answer: 0.10 Type: S   Sect: 3 Obj: 1   RANDOM: Y

30.What is $P(x^2_{(22)} > 36.8)$?

   Answer: 0.025 Type: S   Sect: 3 Obj: 1   RANDOM: Y

31.What is $P(x^2_{(24)} > 36.4)$?

   Answer: 0.05 Type: S   Sect: 3 Obj: 1   RANDOM: Y

32.What is $P(x^2_{(19)} > 36.2)$?

   Answer: 0.01 Type: S   Sect: 3 Obj: 1   RANDOM: Y

33.What is $P(x^2_{(21)} > 29.6)$?

   Answer: 0.10 Type: S   Sect: 3 Obj: 1   RANDOM: Y

34.What is $P(x^2_{(7)} > 16.0)$?

   Answer: 0.025 Type: S   Sect: 3 Obj: 1   RANDOM: Y

35.What is $P(x^2_{(13)} > 27.7)$?

   Answer: 0.01 Type: S   Sect: 3 Obj: 1   RANDOM: Y

36.What is $P(x^2_{(6)} > 12.6)$?

   Answer: 0.05 Type: S   Sect: 3 Obj: 1   RANDOM: Y

37.What is $P(x^2_{(23)} > 32)$?

   Answer: 0.10 Type: S   Sect: 3 Obj: 1   RANDOM: Y

38.What is $P(x^2_{(20)} > 31.4)$?

   Answer: 0.05 Type: S   Sect: 3 Obj: 1   RANDOM: Y

39.What is c so that $P(x^2_{(5)} > c) = 0.005$?

   Answer: 16.8 Type: S   Sect: 3 Obj: 1   RANDOM: Y

40. What is c so that $P(x^2_{(12)} > c) = 0.025$?

   Answer: 23.3 Type: S Sect: 3 Obj: 1   RANDOM: Y

41. What is c so that $P(x^2_{(18)} > c) = 0.01$?

   Answer: 34.8 Type: S Sect: 3 Obj: 1   RANDOM: Y

42. What is c so that $P(x^2_{(21)} > c) = 0.05$?

   Answer: 32.7 Type: S Sect: 3 Obj: 1   RANDOM: Y

43. What is c so that $P(x^2_{(24)} > c) = 0.005$?

   Answer: 45.6 Type: S Sect: 3 Obj: 1   RANDOM: Y

44. What is c so that $P(x^2_{(8)} > c) = 0.025$?

   Answer: 17.5 Type: S Sect: 3 Obj: 1   RANDOM: Y

45. What is the value of c so that $P(x^2_{(9)} > c) = 0.01$?

   Answer: 21.7 Type: S Sect: 3 Obj: 1   RANDOM: Y

46. What is the value of c so that $P(x^2_{(17)} > c) = 0.025$?

   Answer: 30.2 Type: S Sect: 3 Obj: 1   RANDOM: Y

47. What is the value of c so that $P(x^2_{(22)} > c) = 0.05$?

   Answer: 33.9 Type: S Sect: 3 Obj: 1   RANDOM: Y

48. What is the value of c so that $P(x^2_{(7)} > c) = 0.025$?

   Answer: 16.0 Type: S Sect: 3 Obj: 1   RANDOM: Y

49. If the chi-square test statistic is 6.923 from a 3 x 5 table, what decision can we make for $\alpha = 0.025$?

   Answer: We cannot reject H_0, since $6.923 < 17.5$. Type: S Sect: 3 Obj: 2

   RANDOM: Y

50. If the chi-square test statistic is 17.2618 from a 4 x 2 table and $\alpha = 0.01$, what decision can we make?

    Answer: We reject H_0, since 17.2618 > 11.3.   Type: S   Sect: 3 Obj: 2   RANDOM: Y

51. If a contingency table is 4 x 5 and $\alpha = 0.025$, what is the critical value?

    Answer: 23.3 Type: S   Sect: 3 Obj: 2   RANDOM: Y

52. If the chi-square test statistic is 15.26 from a 4 x 5 contingency table with $\alpha = 0.025$, what decision can be made?

    Answer: We cannot reject H_0, since 15.26 < 23.3.   Type: S   Sect: 3 Obj: 2

    RANDOM: Y

53. If the chi-square test statistic is 18.3 from a 5 x 3 contingency table with $\alpha = 0.01$, what decision can be made?

    Answer: We cannot reject H_0, since 18.3 < 20.1.   Type: S   Sect: 3 Obj: 2

    RANDOM: Y

54. If a contingency table is 5 x 5, $\alpha = 0.05$, and the test statistic is 20.52, what decision can be made?

    Answer: We cannot reject H_0, since 20.52 < 26.3.   Type: S   Sect: 3 Obj: 2

    RANDOM: Y

55. If a contingency table is 4 x 4, $\alpha = 0.025$, and the test statistic is 17.82, what decision can be made?

    Answer: We cannot reject H_0, since 17.82 < 19.0.   Type: S   Sect: 3 Obj: 2

    RANDOM: Y

56. If a contingency table is 4 x 3, $\alpha = 0.05$, and the test statistic is 12.1218, what decision can be made?

    Answer: We cannot reject H_0, since 12.1218 < 12.6.   Type: S   Sect: 3 Obj: 2

    RANDOM: Y

57. If a contingency table is 4 x 4, $\alpha = 0.05$, and the chi-square test statistic is 23.56, what decision can be made?

    Answer: We reject H_0, since 23.56 > 16.9. Type: S   Sect: 3 Obj: 2   RANDOM: Y

58. If a contingency table is 4 x 3, $\alpha = 0.025$, and the chi-square test statistic is 16.218, what decision can be made?

    Answer: We reject H_0, since 16.218 > 14.4. Type: S   Sect: 3 Obj: 2   RANDOM: Y

59. If a contingency table is 5 x 2, $\alpha = 0.05$, and the chi-square test statistic is 22.517, what decision can be made?

    Answer: We reject H_0, since 22.517 > 9.5. Type: S   Sect: 3 Obj: 2   RANDOM: Y

60. A cell in a contingency table for a favorite flavor of ice cream and the age group of most patrons has a row total of 60 and a column total of 40, and there are 300 items in the sample. What is the expected frequency of this cell?

    Answer: 8 Type: S   Sect: 2 Obj: 2   RANDOM: Y

61. A cell in a contingency table for a favorite flavor of ice cream and the age group of most patrons has a row total of 250 and a column total of 500, and there are 1,500 items in the sample. What is the expected frequency of this cell?

    Answer: 83.3333 Type: S   Sect: 2 Obj: 2   RANDOM: Y

Use the following table showing alcohol consumption by marital status for the problems below.

Drinks per Month

	0	1-50	Over 50
Single	84	250	86
Married	490	756	154
Divorced	33	74	18

62. What is the expected frequency of the number of singles who have 1-50 drinks per month?

    Answer: 233.2134 Type: S   Sect: 2 Obj: 2   RANDOM: Y

63. What is the expected frequency of the number of married people who have over 50 drinks a month?

    Answer: 185.7069 Type: S   Sect: 2 Obj: 2   RANDOM: Y

64. What is the expected frequency of the number of divorced people who do not drink at all in a month?

    Answer: 39.0103 Type: S   Sect: 2 Obj: 2   RANDOM: Y

65. What is the expected frequency of the number of married people who have 1-50 drinks a month?

    Answer: 777.3779 Type: S   Sect: 2 Obj: 2   RANDOM: Y

66. What is the value that the cell of married people who do not drink at all during the month contributes to the test statistic?

    Answer: 6.4498 Type: S   Sect: 3 Obj: 2   RANDOM: Y

Use the following table for the problems below.

The following table shows averages on a statistic test obtained by students in different majors under 3 different instructors.

	Instructor 1	Instructor 2	Instructor 3
Business	72	73	75
Nursing	82	79	81
Mathematics	85	82	79
Engineering	83	83	82

67. What is the expected frequency of a business major under instructor 2?

    Answer: 72.9498 Type: S   Sect: 2 Obj: 2   RANDOM: Y

68. What is the expected frequency of a mathematics major under instructor 1?

    Answer: 82.8577 Type: S   Sect: 2 Obj: 2   RANDOM: Y

69. What is the expected frequency of a nursing major under instructor 3?

    Answer: 80.2448 Type: S   Sect: 2 Obj: 2   RANDOM: Y

70. What is the value that the cell of an engineering major under instructor 2 contributes to the test statistic?

    Answer: 0.0071 Type: S   Sect: 3 Obj: 2   RANDOM: Y

71. What is the value that the cell of a business major under instructor 1 contributes to the test statistic?

Answer: 0.0595 Type: S   Sect: 3 Obj: 2   RANDOM: Y

Given the dimensions of a contingency table, find the number of degrees of freedom. (Chi-square values rounded to the nearest tenth).

72. a 5 x 6 table

Answer: degrees of freedom = 20 Type: S   Sect: 3 Obj: 2   RANDOM: Y

73. a 4 x 4 table

Answer: degrees of freedom = 9 Type: S   Sect: 3 Obj: 2   RANDOM: Y

74. a 5 x 8 table

Answer: degrees of freedom = 28 Type: S   Sect: 3 Obj: 2   RANDOM: Y

75. a 3 x 4 table

Answer: degrees of freedom = 6 Type: S   Sect: 3 Obj: 2   RANDOM: Y

76. a 5 x 3 table

Answer: degrees of freedom = 8 Type: S   Sect: 3 Obj: 2   RANDOM: Y

77. a 7 x 2 table

Answer: degrees of freedom = 6 Type: S   Sect: 3 Obj: 2   RANDOM: Y

78. a 3 x 3 table

Answer: degrees of freedom = 4 Type: S   Sect: 3 Obj: 2   RANDOM: Y

For the given dimensions of a contingency table, find the appropriate critical value of chi-square at the indicated significance level.

79. a 3 x 5 table, $\alpha = 0.01$

Answer: 20.1 Type: S   Sect: 3 Obj: 2   RANDOM: Y

80. a 2 x 6 table, $\alpha = 0.05$

    Answer: 11.1 Type: S   Sect: 3 Obj: 2   RANDOM: Y

81. a 7 x 6 table, $\alpha = 0.01$

    Answer: 50.9 Type: S   Sect: 3 Obj: 2   RANDOM: Y

82. a 3 x 3 table, $\alpha = 0.05$

    Answer: 9.5 Type: S   Sect: 3 Obj: 2   RANDOM: Y

83. a 7 x 2 table, $\alpha = 0.01$

    Answer: 16.8 Type: S   Sect: 3 Obj: 2   RANDOM: Y

84. a 4 x 2 table, $\alpha = 0.05$

    Answer: 7.8 Type: S   Sect: 3 Obj: 2   RANDOM: Y

85. a 4 x 5 table, $\alpha = 0.01$

    Answer: 26.2 Type: S   Sect: 3 Obj: 2   RANDOM: Y

86. a 6 x 2 table, $\alpha = 0.01$

    Answer: 15.1 Type: S   Sect: 3 Obj: 2   RANDOM: Y

87. A store manager wants to determine if there is a relationship between the day of sale and the price of the item. The following data were obtained.

Price of Item

	Low	Moderate	High
Monday	20	15	4
Tuesday	15	15	6
Wednesday	20	18	6
Thursday	30	25	10
Friday	35	30	14

What is the expected frequency for a high price on Friday?

Answer: 12.0152 Type: S   Sect: 2 Obj: 2   RANDOM: Y

88. A store manager wants to determine if there is a relationship between the day of sale and the price of the item. The following data were obtained.

Price of Item

	Low	Moderate	High
Monday	20	15	4
Tuesday	15	15	6
Wednesday	20	18	6
Thursday	30	25	10
Friday	35	30	14

What is the expected frequency for a low price on Thursday?

Answer: 29.6578 Type: S   Sect: 2 Obj: 2   RANDOM: Y

89. A store manager wants to determine if there is a relationship between the day of sale and the price of the item. The following data were obtained.

Price of Item

	Low	Moderate	High
Monday	20	15	4
Tuesday	15	15	6
Wednesday	20	18	6
Thursday	30	25	10
Friday	35	30	14

What value does the cell representing moderate prices on Wednesday contribute to the test statistic?

Answer: 0.0342 Type: S   Sect: 3 Obj: 2   RANDOM: Y

90. A store manager wants to determine if there is a relationship between the day of sale and the price of the item. The following data were obtained.

Price of Item

	Low	Moderate	High
Monday	20	15	4
Tuesday	15	15	6
Wednesday	20	18	6
Thursday	30	25	10
Friday	35	30	14

What value does the cell representing low prices on Monday contribute to the test statistic?

Answer: 0.2733 Type: S   Sect: 3 Obj: 2   RANDOM: Y

State hypotheses, find the critical value, compute the test statistic, make the decision, and summarize the results for the given data.

91. A survey of 120 congressmen was conducted to determine their opinions on abortion. The following data were obtained:

	Approve	Disapprove
Republicans	15	40
Democrats	35	20
Independents	5	5

At $\alpha = 0.025$, can it be concluded that opinions are related to party affiliations?

Answer:
H_0 : Congressmen's opinions on abortion are independent of political party.
H_1 : Congressmen's opinions on abortion are dependent of political party.
C.V. = 7.4; d.f. = 2
$x^2 = 14.72$
Reject H_0. The opinions of Congressmen are dependent on political party.

Type: S   Sect: 3 Obj: 2   RANDOM: Y

92. A researcher wishes to determine if the age of a person is related to the amount of money the person donates to charity. A sample of 350 people revealed the following data:

Age	Under $500	$500-$1500	$1500-$3000	$3000+
31-40	25	35	10	3
41-50	10	40	60	10
51-60	5	15	40	20
61+	2	10	30	35

At $\alpha = 0.05$, is the amount given to charity independent of the age of the donor?

Answer:
H_0 : The amount given to charity is independent of age.
H_1 : The amount given to charity is dependent on age.
C.V. = 16.9; d.f. = 9
$x^2 = 121.28$
Reject H_0. The amount given to charity is dependent on age.

Type: S   Sect: 3 Obj: 2   RANDOM: Y

93. A small college was interested in determining if there was a relationship between the ages of first-time students and their grade-point averages. Two-hundred students were picked at random and the results of the survey are listed below.

Grade Point Average

Age	Below 2.00	2.0-3.0	3.0-4.0
18-21	30	40	29
41-50	10	20	15
51-60	3	15	20
61+	1	5	12

At $\alpha = 0.05$, can it be concluded that the grade-point average is related to the age of the student?

Answer:
H_0 : A student's grade-point average is independent of age.
H_1 : A student's grade-point average is dependent on age.
C.V. = 12.6; d.f. = 6
$x^2 = 17.92$
Reject H_0. A student's grade-point average is dependent on the student's age.

Type: S   Sect: 3 Obj: 2   RANDOM: Y

94. A math professor felt that there was a relationship between the number of days missed during the semester and the grade that a student received in the course. To test his theory, he compiled the following data:

Days Absent	Pass	Fail
0-3	70	30
4-7	50	15
Over 7	10	25

At $\alpha = 0.025$, can it be concluded that the number of absences is related to a student's final grade?

Answer:
H_0 : The grade in the course is independent of the number of days missed.
H_1 : The grade in the course is dependent on the number of days missed.
C.V. = 7.4; d.f. = 2
$x^2 = 25.57$
Reject H_0. The student's grade is dependent on the number of days missed.

Type: S   Sect: 3 Obj: 2   RANDOM: Y

95. A candidate for president of the United States wants to know whether there are regional differences in the number of supporters he has. A survey is conducted and the results are as follows:

Region	For	Against	Undecided
East	60	40	20
Midwest	40	60	10
West	70	30	20

Test the candidate's claim that his supporters are independent of geographical region at $\alpha = 0.01$.

Answer:
H_0 : Supporters are independent of geographic region.
H_1 : Supporters are dependent on geographic region.
C.V. = 13.3; d.f. = 4
$x^2 = 23$
Reject H_0. The candidate's supporters are dependent on geographic region.

Type: S   Sect: 3 Obj: 2   RANDOM: Y

96. The dean of instruction wanted to determine if the grade distribution
was independent of the subject matter taught. The following table
was compiled:

Subject	A	B	C	D	F
Science	4	12	25	8	12
History	15	20	30	4	5
Psychology	20	40	70	15	20

At $\alpha = 0.01$, can the dean claim that grades are independent of
subject matter?

Answer:
H_0 : Grades are independent of subject matter.
H_1 : Grades are dependent on subject matter.
C.V. = 20.1; d.f. = 8
x^2 = 12.62 or 12.63
Do not reject H_0. Grades are independent of subject matter.

Type: S   Sect: 3 Obj: 2   RANDOM: Y

TRUE/FALSE

97. Nonparametric statistics can be used to test hypotheses other than those involving
population parameters.

Answer: T Type: T   Sect: 1 Obj: 3   RANDOM: Y

98. Nonparametric statistics are more sensitive than their parametric counterparts.

Answer: F Type: T   Sect: 1 Obj: 3   RANDOM: Y

99. Using the data set 5, 8, 12, 7, 6, 15, 6, 9, and 15, the number 8 would have a rank of
2.

Answer: F Type: T   Sect: 6 Obj: 5   RANDOM: Y

100. When using the single sample sign test, if the data value is above the median it is
assigned a "+" sign.

Answer: T Type: T   Sect: 5 Obj: 4   RANDOM: Y

101. The sign test is based on the median and substitutes for a parametric test for the
population mean.

Answer: T Type: T   Sect: 5 Obj: 4   RANDOM: Y

TRUE/FALSE

1. The mean and median are equal when the data is skewed to the right.

   Answer: F Type: T   Sect: 5 Obj: 4   RANDOM: Y

2. The Wilcoxon tests do not consider the ranks of the sample values.

   Answer: F Type: T   Sect: 6 Obj: 5   RANDOM: Y

3. A researcher surveyed married men and single men to ascertain if there was a difference in the number of books each had read during the past year. A sample of 11 married men had a rank sum of 174.5 and a sample of 12 single men had a rank sum of 101.5. The test statistic for married men is 2.62.

   Answer: T Type: T   Sect: 6 Obj: 5   RANDOM: Y

4. The Wilcoxon-Mann-Whitney rank sum test is used to determine if the number of books read by married men is different from the number of books read by single men. If there are 12 married men and 14 single men in the survey, the standard error of the SR values is 378.

   Answer: F Type: T   Sect: 6 Obj: 5   RANDOM: Y

5. When using the Kruskal-Wallis test to compare 3 or more means, it is necessary to consider all data values together and then rank them.

   Answer: T Type: T   Sect: 7 Obj: 6   RANDOM: Y

6. The Kruskal-Wallis test substitutes for a parametric one-way ANOVA.

   Answer: T Type: T   Sect: 7 Obj: 6   RANDOM: Y

7. When the Kruskal-Wallis test is performed and the test statistic H is close to zero, then H_0 will be false.

   Answer: F Type: T   Sect: 7 Obj: 6   RANDOM: Y

8. When using the Kruskal-Wallis test to check between 4 different samples, the H-statistic is 7.92. If $\alpha = 0.05$, we would reject H_0.

   Answer: T Type: T   Sect: 7 Obj: 6   RANDOM: Y

MULTIPLE CHOICE

9. Which of the following is not a nonparametric test?
   a. sign test
   b. Wilcoxon-Mann-Whitney rank sum test
   c. ANOVA
   d. Kruskal-Wallis test

   Answer: c Type: M   Sect: 1 Obj: 1   RANDOM: Y

10. Which of the following is not an advantage of nonparametric statistics?
   a. They are easier to understand.
   b. They can be used when the data are nominal or ordinal in nature.
   c. They can be used to test population parameters when the variable is not normally distributed.
   d. They are harder to understand.

   Answer: d Type: M   Sect: 1 Obj: 1   RANDOM: Y

11. For the data set 8, 7, 17, 18, 13, 14, 15, 13, 8, 7, 6, and 10, what is the rank of the number 13?
   a. 7.5
   b. 5.5
   c. 7
   d. 8

   Answer: a Type: M   Sect: 6 Obj: 4   RANDOM: Y

12. The median score on a 30-point quiz was thought to be 20. A random sample of 18 quizzes yielded the following result:

25	18	25	28	30	18	21	22	15
29	22	16	20	19	16	25	17	19

   What is the test statistic for these data?
   a. 0.24
   b. 0.12
   c. 0.50
   d. 0.03

   Answer: a Type: M   Sect: 5 Obj: 4   RANDOM: Y

13. The median score on a 30-point quiz was thought to be 20. A random sample of 18 quizzes yielded the following results:

    25 18 25 28 30 18 21 22 15
    29 22 16 20 19 16 25 17 19

    If $\alpha = 0.05$, what is the critical value for these data?
    a. 1.96
    b. 1.645
    c. 2.05
    d. 1.28

    Answer: a Type: M   Sect: 4 Obj: 4   RANDOM: Y

14. Eighty teenagers ranked a new CD as above average, below average, or average. The survey results recorded 35 above average, 30 below

    average, and 15 average. What is the value of $\hat{p}$?
    a. 0.4375   b. 0.375   c. 0.50   d. 0.5385

    Answer: d Type: M   Sect: 5 Obj: 4   RANDOM: Y

15. Two independent samples were ranked and a Wilcoxon-Mann-Whitney rank sum test was performed. Sample 1 had 16 items in it with a rank sum of 420. Sample 2 had 19 items in it with a rank sum of 210. What is the test statistic for sample 1?
    a. -2.58
    b. 4.37
    c. 2.58
    d. -4.37

    Answer: b Type: M   Sect: 6 Obj: 5   RANDOM: Y

16. Two independent samples were ranked and a Wilcoxon-Mann-Whitney test was performed. Sample 1 had 15 items in it with a rank sum of 180. Sample 2 had 14 items in it with a rank sum of 255. What is the test statistic for sample 2?
    a. -1.96   b. 1.96   c. 2.07   d. 1.31

    Answer: b Type: M   Sect: 6 Obj: 5   RANDOM: Y

17. The Kruskal-Wallis Test is always
    a. one-tailed left.
    b. one-tailed right.
    c. two-tailed.
    d. many tailed.

    Answer: b Type: M   Sect: 5 Obj: 7   RANDOM: Y

18. Two independent samples were ranked and a Wilcoxon-Mann-Whitney test was performed. Sample 1 contained 15 items, and sample 2 contained 20 items. What is the standard error of the SR values?
    a. 30   b. 900   c. 10.247   d. 29.5804

    Answer: a Type: M   Sect: 6 Obj: 5   RANDOM: Y

19. Two independent samples were ranked and a Wilcoxon-Mann-Whitney test was performed. Sample 1 contained 15 items, and sample 2 contained 20 items. What is μ_{SR} for sample 2?
    a. 270   b. 350   c. 360   d. 262.5

    Answer: c Type: M   Sect: 6 Obj: 5   RANDOM: Y

20. The Kruskal-Wallis test substitutes for which parametric test?
    a. one-way ANOVA
    b. paired difference between 2 means
    c. central location of a population
    d. two-way ANOVA

    Answer: a Type: M   Sect: 7 Obj: 6   RANDOM: Y

21. When finding the critical value for a Kruskal-Wallis test, which distribution must be used?
    a. binomial
    b. t-distribution
    c. chi-square
    d. normal

    Answer: c Type: M   Sect: 7 Obj: 6   RANDOM: Y

SHORT ANSWER

Rank the set of data:

22. 5, 6, 8, 2, 7

    Answer:
    Data: 2 5 6 7 8
    Rank: 1 2 3 4 5
    Type: S   Sect: 6 Obj: 5   RANDOM: Y

23.15, 20, 30, 50, 70

    Answer:
    Data: 15 20 30 50 70
    Rank: 1   2   3   4   5

    Type: S   Sect: 6 Obj: 5   RANDOM: Y

24.8.5, 9, 7.5, 9, 8.5, 7, 9.5, 10

    Answer:
    Data: 7   7.5 8.5 8.5 9   9   9.5 10
    Rank: 1   2   3.5 3.5 5.5 5.5 7   8

    Type: S   Sect: 6 Obj: 5   RANDOM: Y

25.150, 100, 125, 95, 180, 200, 215

    Answer:
    Data: 95 100 125 150 180 200 215
    Rank: 1   2   3   4   5   6   7

    Type: S   Sect: 6 Obj: 5   RANDOM: Y

26.6, 7, 7, 8, 10, 12, 10, 8, 9, 11, 12, 15

    Answer:
    Data: 6   7   7   8   8   9   10   10   11
    Rank: 1   2.5 2.5 4.5 4.5 6   7.5 7.5 9

    Data: 12   12   15
    Rank: 10.5 10.5 12

    Type: S   Sect: 6 Obj: 5   RANDOM: Y

Use the sign test to do the following problem.

27. A real estate agent felt that the median home price in the Park Lane District was $95,000. She selects 20 homes at random and determines their selling prices. The data are listed below (in dollars).

100,000	93,500	110,000	85,000	95,000
96,000	98,000	100,000	89,000	93,000
94,000	95,000	110,000	97,000	90,000
95,000	99,000	92,000	99,000	105,000

What is the test statistic?

Answer: 0.73 Type: S Sect: 5 Obj: 4   RANDOM: Y

28. A real estate agent felt that the median home price in the Park Lane District was $95,000. She selects 20 homes at random and determines their selling prices. The data are listed below (in dollars).

100,000	93,500	110,000	85,000	95,000
96,000	98,000	100,000	89,000	93,000
94,000	95,000	110,000	97,000	90,000
95,000	99,000	92,000	99,000	105,000

At $\alpha = 0.05$, is the real estate agent's claim correct?

Answer:
Do not reject H_0. The real estate agent's claim is correct. The median price of a home in the Park Lane District appears to be $95,000.

Type: S Sect: 5 Obj: 4   RANDOM: Y

Use the steps in hypothesis testing to solve the problem.

29. A Japanese economist believes that the median price of steak in Tokyo is $25. A sample of 18 restaurants yields the following prices for steak:

27	24	29	27	25	30	23	25	24
27	28	35	24	26	25	27	29	30

Test the economist's hypothesis at $\alpha = 0.05$.

Answer:
$H_0 : p = 0.5$; $H_1 : p \neq 0.5$
test statistic = 1.81; C.V. = ± 1.90
Do not reject H_0. The median price of a steak is probably $25.

Type: S Sect: 5 Obj: 4   RANDOM: Y

30. A researcher thinks that the median age of cars owned by students at Paloma High School is 4 years. He takes a sample of 24 cars and finds the following ages:

1	4	3	2	1	5	5	6	1	3	5
3	2	1	1	2	4	5	7	4	3	1
2	3									

    Test the researcher's claim $\alpha = 0.01$.

    Answer:
    $H_0 : p = 0.5$; $H_1 : p \neq 0.5$
    test statistic = -1.96; C.V. = ±2.57
    Do not reject H_0. The median is probably 4 years.

    Type: S   Sect: 5 Obj: 4   RANDOM: Y

31. The owner of a pet shop hypothesizes that the median number of dogs she sells a day is 6. She records the number of dogs sold over the next 21 days. The results are as follows:

4	8	7	3	5	6	9	2	4	5	7
8	8	4	5	6	8	10	5	9	5	

    Test her claim at $\alpha = 0.02$.

    Answer:
    $H_0 : p = 0.5$; $H_1 : p \neq 0.5$
    test statistic = -0.23; C.V. = ±2.33
    Do not reject H_0. The median is probably 6 dogs.

    Type: S   Sect: 5 Obj: 4   RANDOM: Y

32. A weather forecaster believes that the median temperature over the next 20 days will be 75°. He records the temperature over the next 20 days with the following results:

79	82	85	87	89	90	86	78	82	64
69	75	77	75	82	79	78	75	80	90

    Test his claim at $\alpha = 0.05$.

    Answer:
    $H_0 : p = 0.5$; $H_1 : p \neq 0.5$
    test statistic = 3.15; C.V. = ±1.96
    Reject H_0. The median temperature is not 75°; it appears the median is higher than 75°.

    Type: S   Sect: 5 Obj: 4   RANDOM: Y

33. A survey of 200 students showed that 130 were in favor of a smoke-free environment. At $\alpha = 0.05$, test the hypothesis that more than 50% of the students favor a smoke-free environment.

Answer:
$H_0 : p \le 0.5$; $H_1 : p > 0.5$
test statistic = 4.24; C.V. = 1.645
Reject H_0. It appears that more than 50% favor a smoke-free environment.

Type: S   Sect: 5 Obj: 4   RANDOM: Y

34. It is hypothesized that the median number of days for a new roof to be completed is 3. Forty homeowners were questioned, and it was found that 15 homeowners had their roofs completed in less than 3 days, 12 in more than 3 days, and 13 in 3 days. At $\alpha = 0.02$, test the claim that the median time to complete a roof is 3 days.

Answer:
$H_0 : p = 0.5$; $H_1 : p \ne 0.5$
test statistic = -0.58; C.V. = ±2.33
Do not reject H_0. The median time to complete a roof appears to be 3 days.

Type: S   Sect: 5 Obj: 4   RANDOM: Y

35. Test the hypothesis that more than 50% of the employees at Acme Brick Company have worked there more than 15 years, if in a sample of 100 workers, 70 have worked at the company more than 15 years. Use
$\alpha = 0.025$.

Answer:
$H_0 : p \le 0.5$; $H_1 : p > 0.5$
test statistic = 4; C.V. = 1.96
Reject H_0. The median number of years worked at Acme Brick Company is more than 15.

Type: S   Sect: 5 Obj: 4   RANDOM: Y

36. Test the hypothesis that more than 50% of insomniacs take longer than 30 minutes to fall asleep. A study of 200 insomniacs showed 140 took longer than 30 minutes to fall asleep and 60 took less than 30 minutes to fall asleep. Use $\alpha = 0.025$.

Answer:
$H_0 : p \le 0.5$; $H_1 : p > 0.5$
test statistic = 5.65; C.V. = 1.96
We reject H_0. It appears that more than 50% of insomniacs take longer than 30 minutes to fall asleep.

Type: S   Sect: 5 Obj: 4   RANDOM: Y

37. Fifty movie critics rank a new movie as above average, below average, or average. The results were 28 above average, 12 average, and 10 below average. Does this evidence suggest that this movie is going to be popular (above average)? Use $\alpha = 0.05$.

    Answer:
    $H_0 : p \le 0.5$; $H_1 : p > 0.5$
    test statistic = 2.53; C.V. = 1.645
    We reject H_0. It appears that the movie will be popular.

    Type: S   Sect: 5   Obj: 4   RANDOM: Y

38. A teacher hypothesized that the median grade on her next test will be 79. After the test, 18 students scored above 79, 10 scored below 79, and 3 students scored 79. Test her hypothesis at $\alpha = 0.05$.

    Answer:
    $H_0 : p = 0.5$; $H_1 : p \ne 0.5$
    test statistic = 1.51; C.V. = ± 1.96
    Do not reject H_0. It appears the median grade on her test was 79, as she thought.

    Type: S   Sect: 5   Obj: 4   RANDOM: Y

39. A researcher hypothesizes that the median IQ of graduates at West Point is 120. A sample of 120 graduates found 70 with IQs over 120, 40 with IQs below 120, and 10 with IQs of 120. At $\alpha = 0.05$, test the researcher's hypothesis.

    Answer:
    $H_0 : p = 0.5$; $H_1 : p \ne 0.5$
    test statistic = 2.86; critical value = ± 1.96
    We reject H_0. It appears that the median IQ of graduates at West Point is not 120, but perhaps is more than 120.

    Type: S   Sect: 5   Obj: 4   RANDOM: Y

40. Use the following data to answer the question.

5	7	8	10	7	8	7	6	5	4
3	8	10	6	3	2	8	9	7	3

    It was hypothesized that the median is 7. What is the value of $\hat{p}$?

    Answer: 0.4375   Type: S   Sect: 5   Obj: 4   RANDOM: Y

41. Use the following data to answer the question.

5	7	8	10	7	8	7	6	5	4
3	8	10	6	3	2	8	9	7	3

    It was hypothesized that the median was 7. What is the test statistic?

    Answer: -0.5 Type: S   Sect: 5 Obj: 4   RANDOM: Y

42. A teacher who frequently gives 10-point quizzes in his class hypothesized that the median score on a quiz was 6. The next quiz showed 10 students with a score over 6, 4 students with a score below .
    6, and 5 students with a 6. What is the value of $\hat{p}$?

    Answer: 0.7143 Type: S   Sect: 5 Obj: 4   RANDOM: Y

43. A teacher who frequently gives 10-point quizzes in his class hypothesized that the median score on a quiz was 6. The next quiz showed 10 students with a score over 6, 4 students with a score below
    6, and 5 students with a 6. What is the test statistic?

    Answer: 1.60 Type: S   Sect: 5 Obj: 4   RANDOM: Y

44. In testing a hypothesis about the median, the test statistic was 2.13. If we are conducting a two-tailed test and $\alpha = 0.05$, what decision will we make?

    Answer: We reject H_0, since $2.13 > 1.96$. Type: S   Sect: 5 Obj: 4   RANDOM: Y

Use the Wilcoxon-Mann-Whitney Rank Sum Test in the following problem. Assume the samples are independent.

45. An insurance adjustor wanted to know if there was a difference between 2 body shops in their estimates to repair damaged vehicles. He took 11 cars into the two shops and obtained the following estimates.

Shop 1	700	1,000	2,500	950	1,500	850	1,800
Shop 2	750	1,100	2,450	975	1,475	845	2,000

Shop 1	850	1,100	2,000	650
Shop 2	825	1,120	1,975	680

At $\alpha = 0.01$, is there a difference between the two shops?

Answer:
H_0 : There is no difference between the two shops.
H_1 : There is a difference between the two shops.
$z = 0.033$; C.V. $= 2.575$
Do not reject H_0. There is no difference between the two body shops' estimates.

Type: S   Sect: 6   Obj: 5   RANDOM: Y

46. Randomly selected executives are questioned as to the number of hours they work per week. Test the claim that there is no difference between the number of hours a female executive works and the number of hours a male executive works at $\alpha = 0.05$.

Men	45	50	60	55	40	43	52	35	42	50	54
Women	40	45	55	52	42	40	50	51	58	38	44

Answer:
H_0 : There is no difference between the number of hours a female executive and a male executive works per week.
H_1 : There is a difference between hours worked by male and female executives.
$z = 0.36$; C.V. $= 1.96$
Do not reject H_0. There does not appear to be a difference between the number of hours a female executive and a male executive works per week.

Type: S   Sect: 6   Obj: 5   RANDOM: Y

47. An economist hypothesizes that there is no difference between the interest rates charged by banks and the rates charged by savings and loans. He checks the rates at 10 different banks and savings and loans in his area. The results are listed below.

Bank	10	9.5	9.5	10.2	10.3	9.75	9.9
Savings & Loan	10.5	9.75	9.8	10.25	10.2	10	9.8

Bank	10	9.6	9.5
Savings & Loan	10.1	9.75	9.4

At the 0.02 level of significance, test the claim that there is no difference between the two institutions.

Answer:
H_0 : There is no difference between the interest rates the two institutions charged.
H_1 : There is a difference between the interest rates the two institutions charged.
$z = 0.87$; C.V. $= 2.33$
Do not reject H_0. There appears to be no difference between the interest rates the two institutions charged.

Type: S   Sect: 6 Obj: 5   RANDOM: Y

48. Patients are treated with two different drugs to determine if there is a difference in the time (in days) it takes them to recover. Ten patients were randomly selected and their times were recorded.

Drug A	14	15	10	9	13	15	10	11	20	18
Drug B	15	14	11	12	10	14	13	12	18	15

At $\alpha = 0.05$, is there evidence to show that there is a difference in effectiveness between the two drugs?

Answer:
H_0 : There is no difference between the two drugs in amount of recovery time.
H_1 : There is a difference between the two drugs in amount of recovery time.
$z = 0.11$; C.V. $= 1.96$
Do not reject H_0. There appears to be no difference between the two drugs in time it takes a patient to recover.

Type: S   Sect: 6 Obj: 5   RANDOM: Y

49. The Wilcoxon-Mann-Whitney rank sum test is used to determine if the number of books read by married women is different from the number of books read by single women. If there are 20 married women and 17 single women in the survey, what is the standard error of the SR values?

    Answer: 32.8126 Type: S   Sect: 6 Obj: 5   RANDOM: Y

50. The Wilcoxon-Mann-Whitney rank sum test is to be performed. There are 15 items in sample A and 20 items in sample B. What is the standard error of the SR values?

    Answer: 30 Type: S   Sect: 6 Obj: 5   RANDOM: Y

51. The Wilcoxon-Mann-Whitney rank sum test is to be performed. There are 25 items in sample L and 20 items in sample M. What is the standard error of the SR values?

    Answer: 43.7798 Type: S   Sect: 6 Obj: 5   RANDOM: Y

52. The Wilcoxon-Mann-Whitney rank sum test is to be performed on 2 samples. Sample 1 contains 22 items and sample 2 contains 18 items. What is the standard error of the SR values?

    Answer: 36.7831 Type: S   Sect: 6 Obj: 5   RANDOM: Y

53. A Wilcoxon-Mann-Whitney rank sum test is to be performed on two samples. There are 16 items in sample A and 19 items in sample B. The rank sum for sample A is 285 and the rank sum for sample B is 345. What is the test statistic for sample A?

    Answer: -0.0993 Type: S   Sect: 6 Obj: 5   RANDOM: Y

54. A Wilcoxon-Mann-Whitney rank sum test is to be performed. There are 16 items in sample A with a rank sum of 300. There are 19 items in sample B with a rank sum of 330. What is the test statistic for sample B?

    Answer: -0.3974 Type: S   Sect: 6 Obj: 5   RANDOM: Y

55. A Wilcoxon-Mann-Whitney rank sum test is to be performed. There are 12 items in sample M and 14 items in sample N. The rank sum for sample M is 190 and the rank sum for sample N is 161. What is the test statistic for sample M?

    Answer: 1.44 Type: S   Sect: 6 Obj: 5   RANDOM: Y

56. A Wilcoxon-Mann-Whitney rank sum test is to be performed. There are 11 items in sample A with a rank sum of 130. There are 15 items in sample B with a rank sum of 221. What is the test statistic for sample B?

    Answer: 0.96 Type: S   Sect: 6 Obj: 5   RANDOM: Y

57. A Wilcoxon-Mann-Whitney rank sum test is to be performed. There are 16 items in sample X and 20 items in sample Y. What is the mean of the potential SR values, μ_{SR}, for sample X?

Answer: 296 Type: S   Sect: 6 Obj: 5   RANDOM: Y

58. A Wilcoxon-Mann-Whitney rank sum test is to be performed. The test statistic is 2.55, α = 0.05, and a two-tailed test is to be performed. What decision will be made?

Answer: We reject H_0, since 2.55 > 1.96. Type: S   Sect: 6 Obj: 5   RANDOM: Y

59. A Wilcoxon-Mann-Whitney rank sum test is to be performed. Area 1 has 14 items with a rank sum of 153. Area 2 has 16 items with a rank sum of 312. What is the test statistic for Area 2?

Answer: 2.66 Type: S   Sect: 6 Obj: 5   RANDOM: Y

Use the Kruskal-Wallis test.

60. Samples of four different stores' on selected food items are as follows:

Store A	Store B	Store C	Store D
$2.50	$2.75	$2.56	$2.99
1.87	1.78	1.99	2.05
3.95	4.25	3.85	4.50
0.89	0.97	0.89	0.96
1.76	1.65	1.87	1.95

At α = 0.05, is there a difference in prices at the different stores?

Answer:
H_0 : The prices are the same for the four stores.
H_1 : The prices are not the same.
C.V. = 7.8; d.f. = 3
$H = 0.62$
Do not reject H_0. The prices at the four stores appear to be the same.

Type: S   Sect: 7 Obj: 6   RANDOM: Y

61. A store manager wants to see if there is a difference in the average
time (in minutes) a customer has to wait in line for 4 of his
checkers. The times are shown below.

Checker 1	Checker 2	Checker 3	Checker 4
3	3	2	2
5	5	3	3
4	2	5	4
6	3	6	4
7	4	3	5

At $\alpha = 0.10$, is there a difference in the average time a customer has
to wait depending on the checker?

Answer:
H_0 : The average time to wait is the same for all checkers.
H_1 : The waiting time is not the same for all checkers.
C.V. = 6.3; d.f. = 3
$H = 3.051$
Do not reject H_0. It appears that the waiting times are the same for all four checkers.

Type: S   Sect: 7 Obj: 6   RANDOM: Y

62. A botanist plants random samples of each of three different strains
of roses on 6 plots of ground. The plots are the same size and
fertility. The yields per plot are as follows:

Strain 1	Strain 2	Strain 3
12	9	12
10	15	15
9	14	16
11	12	20
10	13	18
13	8	11

At $\alpha = 0.05$, is there a difference in the number of roses produced by
the different strains?

Answer:
H_0 : The number of roses produced by the three strains are the same.
H_1 : The number of roses produced by the three strains are not the same.
C.V. = 5.99; d.f. = 2
$H = 5.547$
Do not reject H_0. It appears that the number of roses produced by the three strains are
the same.

Type: S   Sect: 7 Obj: 6   RANDOM: Y

63. A Christmas tree retailer wants to determine if there is a difference in number of sales between the type of Christmas tree selected. He recorded the number of sales for 5 days and the results were as follows:

<u>Cedar</u>	<u>Pine</u>	<u>Spruce</u>
10	12	14
8	10	12
13	8	10
15	14	15
9	13	12

At $\alpha = 0.05$, is there a difference in the type of Christmas tree purchased?

Answer:
H_0 : The number of Christmas trees purchased by type is the same.
H_1 : The number of Christmas trees purchased by type is different.
C.V. = 5.99; d.f. = 2
$H = 1.005$
Do not reject H_0. It appears that customers have no preference for the type of Christmas tree they purchase.

Type: S   Sect: 7 Obj: 6   RANDOM: Y

64. A carpenter needs to purchase some new tools. He wants to determine if there is any difference in price between 3 different hardware stores in his city. He decides to randomly pick out 7 tools and check their price at each of the 3 stores. The prices (in dollars) are listed below.

<u>Cooper's</u>	<u>Bob's</u>	<u>Big John's</u>
40	42	39
35	34	38
15	14	15
12	13	14
30	28	30
50	55	54
25	24	25

At $\alpha = 0.01$, is there a difference in price between the three stores?

Answer:
H_0 : There is no difference in price between the three stores.
H_1 : There is a difference in price between the three stores.
C.V. = 9.2; d.f. = 2
$H = 0.058$
Do not reject H_0. The three stores appear to have no difference in prices.

Type: S   Sect: 7 Obj: 6   RANDOM: Y

65. A quality control engineer recorded the times (in minutes) required by three workers to assemble a product for the ABC Company. The results are listed below.

Frank	Gary	Wayne
13	12	15
18	16	17
14	13	15
16	15	16
15	14	17

At $\alpha = 0.01$, is there evidence to show that there is a difference in the times of the three workers?

Answer:
H_0 : There is no difference between the three workers in the time required to assemble a product.
H_1 : There is a difference between the workers in the time required to assemble a product.
C.V. = 9.2; d.f. = 2
$H = 3.515$
Do not reject H_0. The appears to be no difference between the workers in the amount of time required to assemble a product.

Type: S   Sect: 7 Obj: 6   RANDOM: Y

66. A researcher wants to determine if there is a difference between the amount of time a person has to wait in line at three banks in town. He records the waiting time (in minutes) of randomly selected people at the three banks. He results are as follows:

National Bank	State Bank	Commerce Bank
5	4.5	3.5
6	3.5	4
5.5	3	3
5	4.5	3.5
5.5	4.5	4

At $\alpha = 0.05$, is there a significant difference between the three banks?

Answer:
H_0 : There is no difference between the three banks in the amount of waiting time.
H_1 : There is a difference between the banks in the amount of waiting time.
C.V. = 5.99; d.f. = 2
$H = 9.875$
Reject H_0. There appears to be a difference in waiting times between the three banks.

Type: S   Sect: 7 Obj: 6   RANDOM: Y

67. Three different methods of teaching history were compared using the Kruskal-Wallis test. Sample 1 contained 8 items with a rank sum of 72. Sample 2 contained 6 items with a rank sum of 56. Sample 3 contained 6 items with a rank sum of 82. What is the test statistic for these data?

    Answer: 2.4667 Type: S   Sect: 7 Obj: 6   RANDOM: Y

68. A sample of four different types of textbooks compares the number of pages in each book using a Kruskal-Wallis test. Sample A has 6 items with a rank sum of 100. Sample B has 5 items with a rank sum of 72. Sample C has 6 items with a rank sum of 86. Sample D has 8 items with a rank sum of 67. What is the test statistic?

    Answer: 5.0263 Type: S   Sect: 7 Obj: 6   RANDOM: Y

69. Three airlines were compared to see if there was a difference in airfares, using a Kruskal-Wallis test. Sample A had 6 items with a rank sum of 60, sample B had 7 items with a rank sum of 85, and sample C had 5 items with a rank sum of 26. What is the test statistic?

    Answer: 5.0120 Type: S   Sect: 7 Obj: 6   RANDOM: Y

70. Four soils were compared to check for a difference in yield, using a Kruskal-Wallis test. Sample A has 6 items with a rank sum of 80, sample B has 6 items with a rank sum of 85, sample C has 5 items with a rank sum of 55, and sample D has 5 items with a rank sum of 33. What is the test statistic?

    Answer: 4.3668 Type: S   Sect: 7 Obj: 6   RANDOM: Y

71. Five brands of televisions were compared to check difference in price, using a Kruskal-Wallis test. Sample A has 6 items with a rank sum of 100, sample B has 5 items with a rank sum of 60, sample C has 6 items with a rank sum of 110, sample D has 5 items with a rank sum of 75, and sample E has 6 items with a rank sum of 61. What is the test statistic?

    Answer: 3.8645 Type: S   Sect: 7 Obj: 6   RANDOM: Y

72. Three drugs were compared to check if there is a difference in effectiveness, using a Kruskal-Wallis test. Sample X has 8 items with a rank sum of 120, sample Y has 6 items with a rank sum of 60, and sample Z has 6 items with a rank sum of 30. What is the test statistic?

    Answer: 9.8571 Type: S   Sect: 7 Obj: 6   RANDOM: Y

73. Four drugs were compared to check if there is a difference in effectiveness, using a Kruskal-Wallis test. Sample A has 6 items with a rank sum of 72, sample B has 8 items with a rank sum of 140, sample C has 6 items with a rank sum of 82, and sample D has 5 items with a rank sum of 31. What is the test statistic?

Answer: 7.4191 Type: S   Sect: 7 Obj: 6   RANDOM: Y

74. Three brands of computers were compared to check if there was a difference in price, using a Kruskal-Wallis test. Sample A has 10 items with a rank sum of 200, sample B has 8 items with a rank sum of 100, and sample C has 9 items with a rank sum of 78. What is the test statistic?

Answer: 10.0635 Type: S   Sect: 7 Obj: 6   RANDOM: Y

75. Four high schools were compared to see if there was a difference in dropout rates between the schools, using a Kruskal-Wallis test. Sample A has 6 items with a rank sum of 112, sample B has 8 items with a rank sum of 140, sample C has 6 items with a rank sum of 96, and sample D has 8 items with a rank sum of 58. What is the test statistic?

Answer: 9.0172 Type: S   Sect: 7 Obj: 6   RANDOM: Y

76. Four college majors were compared to see if there was a difference in starting salaries of their graduates, using a Kruskal-Wallis test. Sample A has 8 items with a rank sum of 160, sample B has 9 items with a rank sum of 210, sample C has 10 items with a rank sum of 250, and sample D has 10 items with a rank sum of 83. What is the test statistic?

Answer: 14.3548 Type: S   Sect: 7 Obj: 6   RANDOM: Y

77. Three types of hair dryers were compared to see if there was a difference in price between the models, using a Kruskal-Wallis test. Sample X had 6 items with a rank sum of 72, sample Y had 8 items with a rank sum of 100, and sample Z had 5 items with a rank sum of 18. What is the test statistic?

Answer: 8.8042 Type: S   Sect: 7 Obj: 6   RANDOM: Y

78. Three colors of cars were compared to see if there was a difference in customer's preference, using a Kruskal-Wallis test. Sample A had 8 items with a rank sum of 120, sample B had 10 items with a rank sum of 150, and sample C had 8 items with a rank sum of 81. What is the test statistic?

Answer: 2.25 Type: S   Sect: 7 Obj: 6   RANDOM: Y

79. A Kruskal-Wallis test is performed to compare 5 different bowling balls. What is the critical value for $\alpha = 0.025$?

Answer: 11.1 Type: S   Sect: 7 Obj: 6   RANDOM: Y

80. A Kruskal-Wallis test is performed to compare 4 different drugs. If $\alpha = 0.05$, what is the critical value?

    Answer: 7.8 Type: S Sect: 7 Obj: 6   RANDOM: Y

81. A Kruskal-Wallis test is performed to compare the grade-point averages of 6 different college majors. What is the critical value if
    $\alpha = 0.025$?

    Answer: 12.8 Type: S Sect: 7 Obj: 6   RANDOM: Y

82. A Kruskal-Wallis test is performed to compare four different brands of VCRs. If $\alpha = 0.01$, what is the critical value?

    Answer: 11.3 Type: S Sect: 7 Obj: 6   RANDOM: Y

83. A Kruskal-Wallis test is performed to compare five different models of cars. If $\alpha = 0.05$, what is the critical value?

    Answer: 9.5 Type: S Sect: 7 Obj: 6   RANDOM: Y

84. A Kruskal-Wallis test is performed to compare six different models of computers. If $\alpha = 0.05$, what is the critical value?

    Answer: 11.1 Type: S Sect: 7 Obj: 6   RANDOM: Y

85. A Kruskal-Wallis test is performed to compare the effectiveness of six different types of medicines. If $\alpha = 0.01$, what is the critical value?

    Answer: 15.1 Type: S Sect: 7 Obj: 6   RANDOM: Y

86. A Kruskal-Wallis test is performed to compare the earned run averages of 8 major league pitchers. What is the critical value if $\alpha = 0.05$?

    Answer: 14.1 Type: S Sect: 7 Obj: 6   RANDOM: Y

87. A Kruskal-Wallis test is performed to compare five different groups on their reaction to a new product. If $\alpha = 0.01$, what is the critical value?

    Answer: 13.3 Type: S Sect: 7 Obj: 6   RANDOM: Y

88. A Kruskal-Wallis test is performed to compare seven different brands of refrigerators. What is the critical value if $\alpha = 0.025$?

    Answer: 14.4 Type: S Sect: 7 Obj: 6   RANDOM: Y

89. A Kruskal-Wallis test is performed to compare 5 different bowling balls. If the test statistic is 6.26 and $\alpha = 0.05$, what decision can be made?

    Answer: Do not reject H_0, since 6.26 < 9.5. Type: S   Sect: 7 Obj: 6   RANDOM: Y

90. A Kruskal-Wallis test is performed to compare 4 different drugs. If the test statistic is 10.2 and $\alpha = 0.025$, what decision can be made?

    Answer: Reject H_0, since 10.2 > 9.3. Type: S   Sect: 7 Obj: 6   RANDOM: Y

91. A Kruskal-Wallis test is performed to compare the grade-point averages of 6 different college majors. If the test statistic is 15.32 and $\alpha = 0.0$, what decision can be made?

    Answer: We reject H_0, since 15.32 > 11.1. Type: S   Sect: 7 Obj: 6   RANDOM: Y

92. A Kruskal-Wallis test is performed to compare 5 different models of cars. If the test statistic is 8.721 and $\alpha = 0.05$, what decision can be made?

    Answer: Do not reject H_0, since 8.721 < 9.5. Type: S   Sect: 7 Obj: 6   RANDOM: Y

93. A Kruskal-Wallis test is used to compare the earned run averages of 8 major league pitchers. If the test statistic is 16.27 and $\alpha = 0.05$, what decision can be made?

    Answer: We reject H_0, since 16.27 > 14.1. Type: S   Sect: 7 Obj: 6   RANDOM: Y

94. A Kruskal-Wallis test is used to compare seven different brands of refrigerators. If the test statistic is 9.35 and $\alpha = 0.025$, what decision can be made?

    Answer: Do not reject H_0, since 9.35 < 14.4. Type: S   Sect: 7 Obj: 6   RANDOM: Y

95. A Kruskal-Wallis test is used to compare six different models of computers. If the test statistic is 11.23 and $\alpha = 0.025$, what decision can be made?

    Answer: Do not reject H_0, since 11.23 < 12.8. Type: S   Sect: 7 Obj: 6   RANDOM: Y

96. A Kruskal-Wallis test is performed to compare the effectiveness of 3 medicines. If the test statistic is 6.57 and $\alpha = 0.025$, what decision can be made?

    Answer: Do not reject H_0, since 6.57 < 7.4. Type: S   Sect: 7 Obj: 6   RANDOM: Y

TRUE/FALSE

```
**************************************************************************
```
The remaining questions in this chapter require the use of the formula generation capabilities of ESATEST III. Please see the ESATEST III documentation if you need directions on how to use them.
```
**************************************************************************
```

97. A contingency table has@4 rows and@3 columns. The degrees of freedom would be `~1*~2`.

    Answer: F,4,2,8,(2,4,6,8),3,3,9,(3,5,7,9) Type: T   RANDOM: F

MULTIPLE CHOICE

98. A chi-square test of independence is to be performed with a contingency table that has@4 rows and@5 columns. How many degrees of freedom are there?
   a. |20
   b. |12
   c. |16
   d. |15

    Answer: (@1-1)*(@2-1),4,2,8,(2,4,6,8),5,3,9,(3,5,7,9) Type: M   RANDOM: F

SHORT ANSWER

99. Given the dimensions of a contingency table, find the number of degrees of freedom. (Chi-square values rounded to the nearest tenth.)
   a@5 x@6 table

    Answer: degrees of freedom = `(~1-1)*(~2-1)`,5,3,9,(3,5,7,9),6,2,8,(2,4,6,8) Type: S

RANDOM: F

MULTIPLE CHOICE

100. For the data set 8, 7, 17, @18, 13, 14, 15, 13, 8, 7,@6, and @10, what is the rank of the number 13?
   a. |7.5
   b. |5.5
   c. |7
   d. |8

    Answer: @1-@1+7.5,18,17,20,0,6,3,6,0,10,10,12,0 Type: M   RANDOM: F

SHORT ANSWER

Rank the set of data:

101.5, 6, @1,@2, 7

   Answer:
   Data: ~2   5   6   7 ~1
   Rank: 1   2   3   4   5,8,10,15,0,2,1,4,0

   Type: S   RANDOM: F

102.Use the following data to answer the question.

   @5   7   8   @10   7   8   7   6   5   4
   @3   8   10   6   3   2   8   @9   7   3

It was hypothesized that the median is 7. What is the value of $\hat{p}$?

Answer: 0.4375,5,1,6,0,10,10,20,0,3,1,6,0,9,10,20,0 Type: S   RANDOM: F